Plants of the Coast Redwood Region

TEXT BY KATHLEEN LYONS AND MARY BETH COONEY-LAZANEO

PHOTOGRAPHY BY HOWARD KING

PUBLISHED BY LOOKING PRESS
21600 BIG BASIN WAY #5, BOULDER CREEK, CA 95006

PRINTED IN HONG KONG THROUGH
THE PRINTING CONNECTION, BERKELEY, CA

Plants
of the
Coast
Redwood
Region

PEN AND INK ILLUSTRATIONS BY EMILY SILVER
BOOK DESIGN AND PRODUCTION BY STEPHEN KOWALSKI

DEDICATION

Through the publication of this second edition of "Plants of the Coast Redwood Region" we pay tribute to the artistry and dedication of Howard King.

His photographs of the coast redwoods and the flora of the natural community which surround these giants have inspired many to exert greater effort toward redwood preservation. Howard's documentation of the coast redwood began with his early hikes to Big Basin Redwoods and Butano State Parks. These photographic renderings encouraged public support for the parks and their subsequent protection and expansion. "His photographs are the mirror through which thousands of people view the redwoods" wrote Tony Look, Founder and Director of the Sempervirens Fund.

This edition, in a small way, is intended to let Howard know we appreciate his efforts in the preservation of the coast redwoods—the *Sequoia sempervirens.*

His friends and admirers

July, 1988

ACKNOWLEDGEMENTS

We greatly appreciate the support and encouragement of the many people who helped make this 2nd edition possible.

Barb Danforth, for her review of the original manuscript, her companionship along the trails, and preliminary plant sketeches; Jim and Juanita Cooney for their help on hiking, typing, and pulling the book together; Neil and Marge Lyons, for their continued encouragement; Andy Lazaneo, for his patience and incredible babysitting talents; and to Gary Kohler, for his constant companionship.

We are also grateful to Looking Press for publishing this 2nd edition and to the two principals—Howard King for his wonderful photographs and never-ending enthusiasm for the redwoods, and Tony Look for his determination to secure the land necessary to complete the boundary of Big Basin Redwoods State Park. To the rangers, park staff, and the Natural History Associations of the redwood region, California State Park System, we give our heartfelt thanks.

For their adherence to perfection and encouragement we wish to thank Stephen Kowalski Designworks, Berkeley, California and Meredith Hoffman, Humanitech, Palo Alto, California.

We expressly want to thank Save-the-Redwoods League for their support and use of the cover photograph. And, finally, to Sempervirens Fund for their work in the preservation of the redwoods of Big Basin Redwoods State Park.

Metric System Table

1 mm. = approx. 1/25 of an inch

10 mm. = 1 cm. (approx. 2/5 of an inch)

10 cm. = 1 dm. (approx. 4 inches)

10 dm. = 1 m. (approx. 40 inches)

Contents

INTRODUCTION

This book is designed for hikers and nature lovers who are interested in the plants around them, but have little or no botanical experience. Each plant has a color photograph and a short non-technical description, which includes general localities where the plant can be found and some interesting historical and cultural information. In the few places where specialized botanical vocabulary was necessary for accuracy the terms are explained in the Glossary. Associated with the Glossary are drawings of flower parts and other plant structures.

Although most of the field work for this book was done in the Santa Cruz Mountains and especially in Big Basin Redwoods State Park, the majority of these plants grow throughout the coastal redwood region. This region extends along the coastal mountains from Santa Cruz north to the Oregon border.

The coast range of northern California contains a great diversity of plant and animal communities. Many of these communities merge with others to at least some extent. Probably the best known, and the basis of this book, is the redwood forest. Coast redwoods form almost pure stands with only occasional Douglas fir, tan oak, madrone, and bay trees intermingled. In the northernmost portion of its range are the addition of Sitka spruce and coast hemlock. The moist, shady forest floor covered with thick, acidic redwood duff often restricts the diversity of plants that can grow beneath these trees. Shrubs such as huckleberry, western azalea, and California rhododendron can tolerate this shade and sometimes create dense thickets. Streams in the redwood forest often have a lush growth of herbaceous plants but are so dominated by the towering redwoods that they are not considered true riparian or streamside communities.

Mixed evergreen forests are often found on drier slopes adjacent to redwood forests. Douglas fir and tan oak are abundant with smaller concentrations of oak, madrone, bay, wax myrtle, hazel, vine maple, and redwood. Since these forests are much more open, allowing more light penetration, many species of plants grow within this forest. Most plants in this book can be found here.

Oak woodland communities grow along many of the dry ridges. These forests, comprised of coast live, canyon, interior live, or black oak, are usually open, with large sunny patches of grasses and flowers scattered between the trees. Oak woodlands are known for their great spring wildflower displays.

On the hottest, driest, southwest-facing slopes are the dense impenetrable thickets of the chaparral community. The plants in these areas survive extremely harsh environmental conditions and have developed several common adaptations. Most plants are evergreen with small, light-colored, moisture-conserving leaves. Often the leaves are covered with small hairs or a waxy coating for additional protection from moisture loss. Most plant species are also adapted to frequent fires and are able to stump sprout following a fire. The most common plants in this community are manzanita, chamise, yerba santa, sage, and buckbrush. Knobcone pine is sometimes included as a member of the chaparral and sometimes as a separate community.

At the opposite end of the environmental scene is the riparian or streamside community. Plants of this habitat are not only assured a constant water supply but are often flooded in winter. Because of their limited distribution within California and high value to native wildlife, riparian communities are extremely important habitats. Sycamore, maple, box elder, willow and cottonwood dominate the vegetation.

GUIDELINES FOR USING THIS BOOK

The book is arranged into five major categories: conifer, broadleaf trees, flowers, ferns and exotic (non-native) plants. Conifer trees are those which bear their seeds in cones; they are usually evergreen and have needle-like leaves. Broadleaf trees reproduce by flowers, some so small that they are unnoticeable on casual observation. Their broad, flat green leaves can be evergreen or deciduous.

The flower section of this book includes both shrubs and herbaceous (non-woody) plants. For easy identification these plants are usually divided into several color groups: pink-red, blue-purple, green-white, and yellow-orange. These colors are not absolute, however, and in some cases more than one color group should be checked. For example, some plants have shades ranging from pale pink to purple depending upon the individual plant and its stage of maturity. A few plants vary drastically from white to purple as they mature.

Ferns and fern allies are also included. These are herbaceous plants which reproduce by spores.

The last section of the book contains exotic (non-native) plant species. These plants are those which have been introduced into California during the past two hundred years from other locations and have become established within our native plant communities. Because of their weedy nature and their tendency to out-compete the native flora, several areas have active eradication programs for these species.

Each plant has its scientific name listed as well as its common name. Unlike common names, which can vary from region to region and person to person, scientific names are universally accepted names. The first word, the genus name, is capitalized and may be shared by many similar plants. The second part, in lower case, is the specific name, which only belongs to that species. Both of these names are always underlined or italicized. Occasionally, varieties are acknowledged within a certain species. Also included are the common and scientific family names; a family is composed of several different genera. The most common blooming period for each plant is listed.

We would like to remind the reader that although many plant uses are listed, it is illegal to pick anything from California State Park lands. Without this law, many of our beautiful plants would no longer be here for us or future generations to enjoy. We also take no responsibility for the usefulness or edibility of any of the plants described. Individual palates differ and what may be acceptable to one may not be to another.

Most of the plants are forage for native wildlife. Wildlife species of this area generally have plenty of natural food available to them. Please do not feed them; human handouts are detrimental to their health.

LOCAL CONSERVATION ORGANIZATIONS

We hope that this book increases your enjoyment and awareness of plants of the coast redwood region. Should you be interested in learning more about them or taking an active part in their preservation, there are several fine organizations which are dedicated to these goals. The Sempervirens Fund and Save-the-Redwoods League are conservation organizations which raise funds for the enlargement of redwood parks. Other organizations include State and National Park cooperating associations (natural history and docent activities) and local chapters of the California Native Plant Society, Sierra Club, California Nature Conservancy, and Audubon Society.

CONIFERS

COAST REDWOOD

Taxodiaceae
(Redwood) Family

Sequoia sempervirens
Blooms: May-July

The coast redwood is the most outstanding tree in this area and because of its immense size and beauty, it was instrumental to the establishment of Big Basin Redwoods State Park in 1902 and the northern California redwood parks in the following years.

Often reaching a height of 300 to 350 feet and a diameter of 12 to 16 feet, the coast redwood is the largest tree along the Pacific Coast. The thick bark, with its deep furrows running the length of the tree, is a rich reddish brown. It is this bark that gives the redwood its excellent fire resistant quality. The dark green leaves are needle-like and grow flat off the branches.

Small cones, usually about an inch long, hang from the branch tips, releasing tiny brown seeds when mature. Of these seeds, only 15 to 20 percent germinate and grow into seedlings. Redwoods are also capable of sprouting from the roots of parent trees. These sprouts, because of already established root systems, grow more vigorously than seedlings and so are the more common form of reproduction.

Coastal Indians were the first to make use of the stately redwood. The strong roots were dug up, stripped for their fibers, and used as a thread in baskets. Today, the coast redwood is primarily used for lumber.

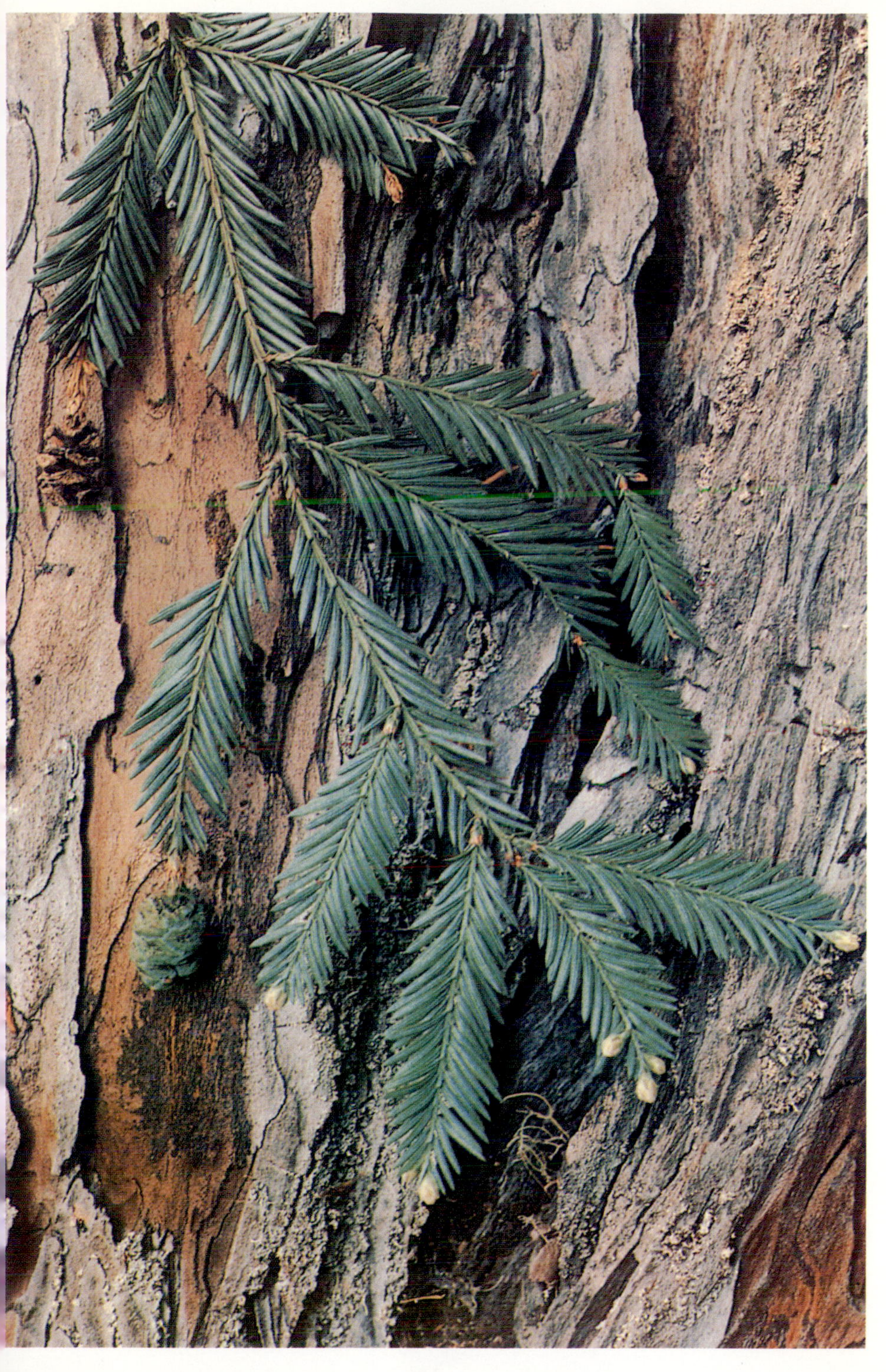

DOUGLAS FIR

Pseudotsuga menziesii

Pinaceae
(Pine) Family

Douglas fir is one of the most common trees in the mixed evergreen forest, and is a prominent member of the redwood forest.

Besides being one of the most common trees, this native is one of the largest, second in size only to the coast redwoods. Because of their size, Douglas firs are sometimes confused with redwoods, but there are many distinguishing features between the two trees, the most outstanding being the bark. Both trees have furrowed bark, but the redwoods have long deep parallel grooves running the entire length of the tree, while Douglas-fir grooves are short and not so symmetrically parallel. The bark of Douglas fir is a dark gray, while redwood bark is a deep reddish brown.

The light green single needles grow in whorls around drooping stems. Althought these stems droop, the sharply uplifted branches give the tree an erect profile. The cones, which mature in one year, hang from the branch tips. These cones are 3 to 4 inches long and have 3-forked bracts between the rounded scales.

Douglas fir had many uses in the past, and is still important today. Medicinally it was used as a treatment for rheumatism and tuberculosis. For rheumatism, Indians covered boughs with soil and burned them in the steamhouse while the patient lay on a blanket above. A tea, rich in vitamin C, was brewed from the needles to treat lung trouble and tuberculosis.

Smoke from burning limbs was used as a good luck charm. It was believed that by holding their bows over the smoke, hunters would be undetectable to deer, and the hunt would be successful. Another hunting use involved fashioning a shaft from the branches, which was used as a salmon spear.

In basketry, the long thin roots were separated and used for thread. Today Douglas fir is an important tree to the lumber industry, and is the most important lumber tree from California to British Columbia.

Douglas Fir

Douglas Fir

COAST HEMLOCK

Tsuga heterophylla

Pinaceae
(Pine) Family

This tree grows in many locations near the coast in the northwestern California area. Further north towards Alaska the hemlock forms dense, almost pure stands.

A large tree with deeply furrowed brown bark, the coast hemlock can reach heights of 100 to 200 feet. Long roots often entwine themselves over nearby fallen logs, giving it the nickname "Octopus tree." Its evergreen, needle-like leaves are spirally arranged along the slender, drooping branches, but may appear flattened. The leaves, dark green above and whitish beneath, are twisted near the base and attach to the branch with short, peg-like petioles. They range from 1/4 to 3/4 inch in length. The solitary cones are cylindrical, 1/2 to 1 inch long, and hang from the ends of the branchlets. Tiny winged seeds mature in their first year and are released to the wind for dispersal. The wood from the coast hemlock is used in the construction industry and is known for its durability.

GRAND FIR

Abies grandis

Pinaceae
(Pine) Family

This forest tree is native to the coastal mountains from northern California to British Columbia.

Reaching 75 to 200 feet in height, the grand fir is an evergreen tree. Brownish bark, thick and furrowed, covers the trunk and long drooping branches. The leaves, dark green above and whitish beneath, form flat sprays from the branchlets. They are 3/4 to 2 inches long and have a characteristic rounded notch at the tip. These "true firs" leave a circular scar on the branch when the leaf falls. The cones, comprised of many spirally arranged papery scales, sit erect on the upper side of the branches. These become mature in the first year, releasing scales and long-winged seeds into the wind.

Not being a durable wood, the grand fir is mostly used to make boxes. Its seeds, however, are an important food source for native wildlife.

Coast Hemlock

Grand Fir

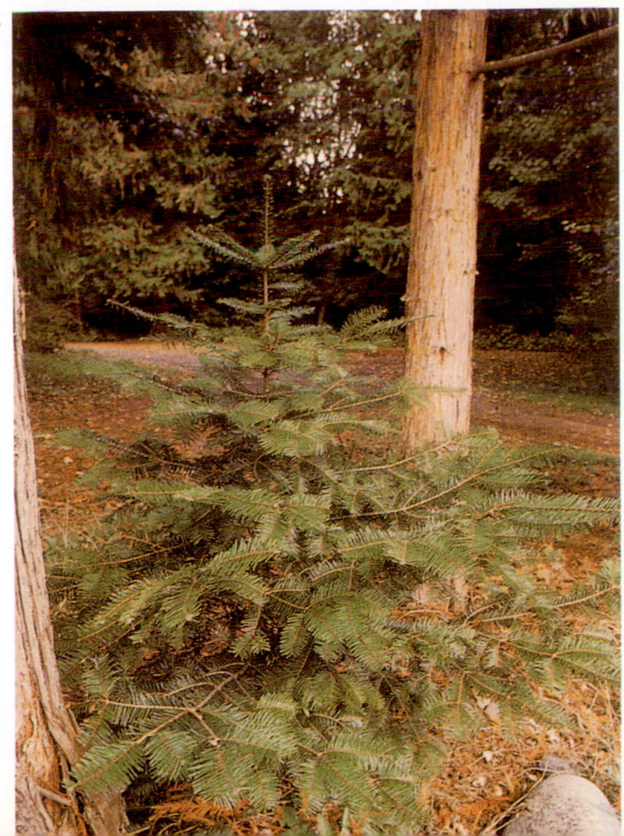

SITKA SPRUCE

Picea sitchensis

Pinaceae
(Pine) Family

The Sitka spruce is a stately tree inhabiting the coastal mountains from northern California to Alaska. It thrives in the cool, moist environment of this area, along with hundreds of clinging mosses and ferns which often cover the tree's buttressing trunk.

A tall forest tree, this spruce can reach heights of 125 to 200 feet. Its gracefully whorled branches and lush growth give it a conical shape. The bark, when visible, is a beautiful reddish-brown with thin, loose scales. The flattened leaves, whitish on the upper surface and bright green beneath, are simple and arranged spirally along the stem. Jointed at the base, they are 1/2 to 1 inch long with sharply pointed tips. The brown cones, composed of several spirally arranged scales, are 2 to 4 inches long. Two seeds are at the base of each scale.

The spruces can be distinguished from the firs and false-hemlocks by the rough texture of the branches. This is due to the small woody pegs left on the branch after a leaf falls.

During World War I the Sitka spruce was extensively harvested; the lumber was used to build wood-framed, canvas-covered fighter planes. Today, this tree is often planted as an ornamental in Oregon and Washington states.

Sitka Spruce

Sitka Spruce

SANTA CRUZ CYPRESS

Cupressus abramsiana

Cupressaceae
(Cypress) Family

The Santa Cruz cypress, found in only a few locations in the Santa Cruz Mountains, is on the rare and endangered list. It inhabits dry, often sterile, inland marine sand deposits and sandstone outcroppings. Unfortunately, due to increased development within this unique environment, the cypress is threatened with extinction.

Related to many cultivated varieties of cypress, this native grows to heights of 50 to 60 feet and has thick bright green foliage. Small overlapping scale-like leaves grow along the branches. The cones, found at the tips of these branches, are closed until the second season and upon maturity release tiny, brown, winged seeds.

CALIFORNIA NUTMEG

Torreya californica

Taxaceae
(Yew) Family

The California nutmeg grows in many habitats throughout the Coast Range but is not particularly common in any.

Although the California nutmeg is most often seen in a shrub-like form, it occasionally grows forty or more feet tall. Lying flat off the branches, the needles are a glossy green with sharp pointed tips. The large fruit, formed after pollination of the female ovules, hangs from the outer branches, changing from olive green to a deep purple as it ripens. A large seed inside somewhat resembles commercial nutmeg, but it is not related.

Early Indian tribes used the root of the California nutmeg for a basketry thread. The sharp needle tips were also used as instruments for tattooing.

Santa Cruz Cypress

Santa Cruz Cypress

California Nutmeg

KNOBCONE PINE

Pinus attenuata

Pinaceae
(Pine) Family

The knobcone pine is native to California and can be found on dry slopes, often where few other trees can survive. This adaptation to poor soil conditions makes it a hardy pioneer plant for recently burned or disturbed areas.

It grows from Santa Cruz county, north to Del Norte and Siskiyou counties.

Because of its harsh environment, the knobcone pine is usually sparse and scraggly. The needles, which grow in clusters of 3 on slender branches, range from 4 to 7 inches in length. Closed woody cones adhere tightly to the trunk and branches, opening only in extreme heat to release the small seeds.

Like most other pines, the seeds are edible either raw or roasted. They can be gathered by heating the cones until the bracts open, allowing the seeds to fall out.

Knobcone Pine

Knobcone Pine

BROADLEAF TREES

MADRONE

Ericaceae
(Heath)Family

Arbutus menziesii
Blooms: March-May

Although only occasionally found in the deeper redwood valleys, the madrone is quite common on the upper slopes and ridges in the mixed evergreen forest.

Because of its constant search for light, this large native tree often assumes unusual gnarled and twisted shapes. Another unusual feature of the tree is the characteristic thin bark on the trunks and limbs. Each summer much of this outer bark peels off and hangs in tatters, exposing smooth light green wood which weathers to a rich red brown.

The evergreen leaves are a waxy green and grow alternately along the branches. In early spring small clusters of waxy, white, bell-shaped flowers appear, and in late summer these mature into large orange berries.

This tree was used for food, medicine, and utensils by both Indians and settlers. Roots and leaves were brewed into a tea to treat stomach aches. Sores and wounds were treated with a lotion made from the leaves and bark. Whether eaten raw or boiled in baskets with hot rocks, the berries were an important food. When dried they were stored and used as an important winter staple.

The fine-grained wood of the madrone was used by Indians for lodgepoles and by some early settlers to make stirrups. Charcoal from the burned wood reportedly made an excellent gunpowder, which was sold commercially.

Madrone

Madrone

Madrone

OAKS

Quercus species

Fagaceae
(Oak) Family

Although oaks have variable forms, they have several common characteristics. Ripening from the female flowers are the acorns, which are composed of a smooth, thin-shelled nut protruding from a scaly cap. The male flowers hang down in pendant catkins.

USES: The oaks were one of the most important food sources of the California Indians. Acorns were gathered by the women and thrown over their backs into large baskets, called burden baskets. Older women didn't join in the gathering, but instead enjoyed the privilege of sorting the wormy acorns from the whole ones. After years of eating acorns containing grit from leaching and grinding with rocks, these women had only stubby remnants of teeth. Therefore, the soft worms from the "bad" acorns were considered a great delicacy. This privilege of the old served a two-fold purpose, since it also kept worms from laying eggs in the acorns and destroying the meat before it could be used.

After sorting, the acorns were soaked overnight in hot water, hulled, and ground into meal. Since acorns contain a bitter tannic acid, they had to go through a complicated process of leaching before being eaten. This was usually done either by placing the meal in a frame of twigs and letting water run over it, or by boiling it by the hot rock method. In this process the meal was placed in watertight baskets and hot rocks from the fire were added until the water boiled. Since the water had to be changed several times to remove the dissolved tannic acid, this was a tedious process. After leaching, the meal was made into a thin gruel or wrapped in fern leaves and baked on hot rocks.

Acorns were important medicinally as well. Before it had been discovered by modern civilization, a type of penicillin was used by the Indians to draw out sores and boils. The acorn meal was covered tightly to cause mold to form, then, when this layer of mold or "skin" was strong enough to be pulled off, it was rolled into sheets and stored until needed.

The wood from some of our local oaks has been important in the past, and some is still used today. Although the spreading form of the coastal live oak precludes its use for lumber, the wood burns well and is often used as a fuel and a source of charcoal. The maul oak (canyon live oak) received its name because its wood is so hard and heavy that it was used to make maul heads in the pioneer days.

Coast Live Oak

COAST LIVE OAK

Fagaceae
(Oak) Family

Quercus agrifolia
Blooms: February-April

The coast live oak is the most common oak in these mountains, inhabiting mixed evergreen forests, oak woodlands, and grasslands.

When uncrowded by other trees, this oak develops a distinctive rounded crown with wide spreading branches. The sharply-toothed leaves have a shiny dark green upper surface and a lower surface covered with tufts of tan fuzz. Sometimes mistaken for the interior live oak, the coast live oak has deep curved-under leaves. Oblong acorns grow from a fringed cap and end in a pointed tip.

CALIFORNIA BLACK OAK

Fagaceae
(Oak) Family

Quercus kelloggii
Blooms: March-May

Growing in open oak woodlands, black oaks are most often seen at higher elevations on the eastern slopes of the mountains. Several large groves are present at Castle Rock State Park in the Santa Cruz Mountains.

Unlike most other oaks, the black oak is deciduous. When the leaves first appear in the spring they are red and covered with fine hairs. Later they mature to deep green. Like those of the white oak (*Quercus lobata*), which grows in the interior valleys, these leaves are deeply lobed. However, they can be distinguished by the spiny points on the lobe margins.

Coast Live Oak

Coast Live Oak

California Black Oak

INTERIOR LIVE OAK

Fagaceae
(Oak) Family

Quercus wislizenii
Blooms: April-May

This medium-sized oak is located on dry wooded slopes, often hybridizing with the coast live oak.

The glossy green leaves of the interior live oak are leather-like with sharply-toothed margins. They differ from the coast live oak in that they are relatively flat and lack the short fuzz on the under surface. Like canyon live oak, the slender acorns mature and drop in their second autumn.

CANYON LIVE OAK

Fagaceae
(Oak) Family

Quercus chrysolepis
Blooms: May-June

The canyon or maul oak, like most oaks, is found on dry wooded slopes of the Coast Range.

Largest of the western oaks, this oak has been known to grow to a diameter of ten feet. The leaves have variable margins, ranging from smooth to coarsely-toothed, often on the same twig. They are usually characterized by a golden fuzz on the underside and a glossy green upper surface. In autumn, the second year acorns, with the golden-tinged rounded caps, fall to the ground.

SCRUB OAK

Fagaceae
(Oak) Family

Quercus dumosa
Blooms: April-June

Scrub oak is found growing in the coastal mountains, usually in the dry chaparral regions.

Due to the harsh conditions in which it is found, this oak grows as a shrub. The glossy green leaves are small, thick, and leathery.

Interior Live Oak

Canyon Live Oak

CALIFORNIA BAY

Lauraceae
(Laurel) Family

Umbellularia californica
Blooms: December-April

California bay is common on wooded slopes of the Coast Range. It reaches its largest size in northwestern California and southwestern Oregon, where it is know as Oregon myrtle.

The leaves, which are wedge-shaped at the base and pointed at the tips, somewhat resemble wax myrtle leaves, but can be distinguished by their pungent fragrance. Arranged in an alternate pattern, these leaves are leathery and usually dark green. In early spring, small yellowish-green flowers develop and profusely cover the tree.

The bay tree had many uses for the Indians, the most important being medicinal. Leaves were used to cleanse wounds and to cure headaches. For headaches the leaves were either bound around the head or placed up the nostril. Leaves steeped in boiling water were used as a disinfectant, while smoke from leaves burned directly on the fire was used as a "vaporizer" for colds. For treatment of rheumatism the Indians would rub their bodies with bay oil while taking their steambaths. The combination of the stinging bay oil and a complete body rubdown was supposed to be an effective cure. Evidently there was some merit to this, since the settlers later adopted the treatment. However, instead of using it in steambaths, they mixed the oil with lard and covered their bodies with liniment.

Bay nuts were roasted and later eaten whole, or ground into flour and baked into breads. Roasting removed the bitter taste caused by the high acid content.

A more curious use for the leaves was as a flea repellent in Indian dwellings. By spreading leaves on the floor they kept their living quarters flea free.

Today the bay is still useful. Leaves are dried and ground into a spice which is similar to the expensive European bay leaves. The wood, which is white and fine-grained, is used under the name of Oregon myrtle to make bowls and other ornamental objects.

California Bay

BLACK COTTONWOOD

Salicaceae
(Willow) Family

Populus trichocarpa
Blooms: February-April

Black cottonwood is found in riparian communities, along the coast, from southern California to Alaska.

Commonly reaching heights of 75 feet or more, the cottonwood tree is known by its gray striated bark. Smooth green spear-shaped leaves hang from the branches on long stems. In early spring, male and female flowers, arranged in catkins, hang from the branches.

Both the catkins and inner bark of cottonwood are edible, either raw or boiled.

WESTERN SYCAMORE

Platanaceae
(Sycamore) Family

Platanus racemosa
Blooms: March-April

Sycamores grow along river banks throughout the central Coast Range, often forming large groves.

Reaching heights of up to 100 feet, this tree has smooth, thin, whitish bark, arranged in jigsaw-like pieces. The sharp contrast of the white tree trunk with the surrounding vegetation makes it a striking tree.

The palmate leaves of the sycamore are light green and deeply lobed. Small hairs cover both surfaces, giving them a velvety texture. Small clusters of flowers hang from the stems in spring, later ripening into round fruits which remain on the tree throughout winter, then break open to release small winged seeds.

The wood of the sycamore is strong, fine-grained, and hard to split; it is only used as a fuel source.

Black Cottonwood

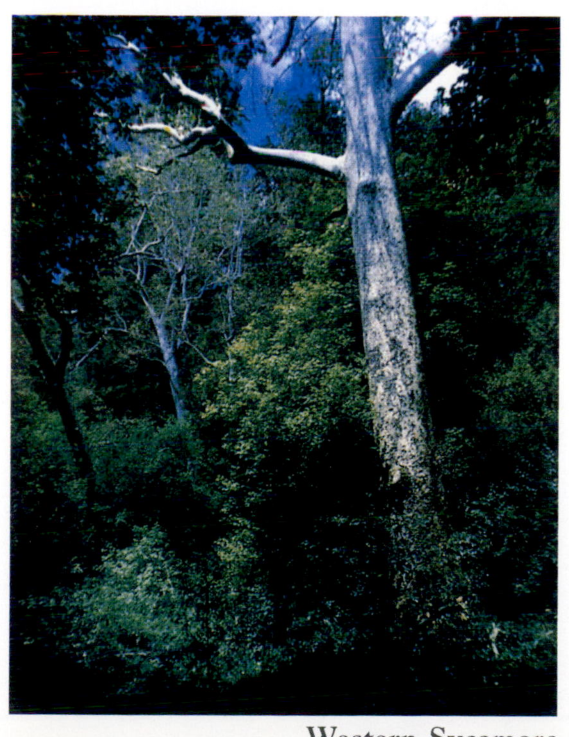

Western Sycamore

RED ALDER

Betulaceae
(Birch) Family

Alnus rubra
Blooms: (female) January-March

Red alder grows along streams throughout the Coast Range from Santa Cruz county to Alaska.

This native tree can be recognized by its gray-white patchy bark and coarsely toothed alternate leaves. Although similar to the white alder in appearance, the leaf edges of red alder curl under, while the leaf edges of white alder do not. In February and March, long male catkins release pollen which fertilizes the cone-like female catkins. By autumn, the seeds are ripe and are shed from the cone as winged nutlets.

The main use for red alder by California Indians was in basketry. Roots were used for weaving thread, while the inner bark produced an orange dye. Chunks of the bark were cooked until soft, chewed until the dye was released, then spit into a stone dish. The acid in the saliva set the color, which turned to a rich brown with age.

BIG LEAVED MAPLE

Aceraceae
(Maple) Family

Acer macrophyllum
Blooms: March-May

This large native tree grows along streams and moist areas throughout the coastal mountains, often forming dense stands.

When the maple is young it has smooth light gray bark, but as it grows older it becomes darker gray-brown with large cracks and ridges. At maturity the tree can reach up to 80 feet in height. Its large, 3 to 5-lobed, palmate leaves turn to bright yellow or orange in the fall. Before new leaves grow again in spring, long drooping clusters of yellowish-green flowers appear. These flowers ripen into double-winged fruits, called samaras, which blow easily in the wind.

Although this maple is not widely used, its sap can be made into syrup.

Red Alder

Big Leaved Maple

Big Leaved Maple

CALIFORNIA WAX MYRTLE

Myricaceae
(Wax Myrtle) Family

Myrica californica
Blooms: April-June

Wax myrtle is found scattered throughout the redwood and mixed evergreen forests.

Growing as a large shrub or small tree, depending on its environment, myrtle is sometimes confused with California bay because of similar leaves. Like the bay, the leaves are long, slender, and dark green, but they differ in that they are slightly serrated and somewhat broader at the tip. Also, they lack the pungent odor of the bay. The short male catkins are borne on lower leaf axils, while the longer female catkins are borne on upper axils. Berries, produced by the female catkins, are greenish purple with a waxy white covering.

Although there is no reference to the western myrtles being used this way, the waxy berries of an Atlantic species are used to make candles.

TAN OAK

Fagaceae
(Oak) Family

Lithocarpus densiflorus
Blooms: May-June

The tan oak is a common resident of the redwood and mixed evergreen forests and can be found on most forested trails in these mountains.

This tree, which is not a true oak, grows straight and tall and has smooth gray bark. Its glossy green leaves are brittle and coarsely toothed along the margins. In late spring erect clusters of flowers appear, which mature into acorns after two years. These differ from true oak acorns by the bristly appearance of the cap.

Tan oak acorns were prepared like oak acorns for food and medicine (see oaks). Tannic acid, used to tan leather, is derived from these trees. At one time Big Basin Redwoods State Park was an important source of tan oaks. Large sections of bark were stripped from the trees, dried, and shipped to Santa Cruz by wagon. There they were boiled to leach the acid from the bark. Northern California also supported the tanning industry by the tan oak forests of Humboldt and Del Norte counties.

California Wax Myrtle

Tan Oak

VINE MAPLE

Aceraceae
(Maple) Family

Acer circinatum
Blooms: April-May

This deciduous tree inhabits streambanks and moist canyons of mixed evergreen and redwood forests from Mendocino county in northern California all the way to Alaska.

Growing anywhere from 5 to 35 feet in height, the vine maple can be an erect tree or a nearly prostrate creeping vine. Its delicate green leaves, 2 to 6 inches wide, have 7 to 10 rounded lobes with small, sharply toothed edges. The petioles, that attach the leafblade to the branch, are 1 to 2 inches long. Since this tree is deciduous, the leaves turn a brilliant orange-red in autumn and then fall to the ground. The small flowers are reddish-purple and hang in clusters from the branches. These mature into reddish fruits, which release the 2-winged samaras, or seeds.

This species has less deeply lobed leaves than the big leaved maple.

CALIFORNIA BOX ELDER

Aceraceae
(Maple) Family

Acer negundo ssp.*californicum*
Blooms: March-April

Box elder grows along the banks of streams. It is common throughout the entire Coast Range.

Reaching 30 or more feet in height, this tree has opposite compound leaves that hang on long slender stems. Each leaf is composed of 3 leaflets that are lobed and have toothed margins. In spring, small male and female flowers appear on separate trees. The male flowers hang on small thin stems, profusely covering the trees. Winged seeds called samaras are red when young, turning a golden straw color with age.

Vine Maple

California
Box Elder

COAST RED ELDERBERRY

Caprifoliaceae
(Honeysuckle) Family

Sambucus callicarpa
Blooms: March-June

BLUE ELDERBERRY

Caprifoliaceae
(Honeysuckle) Family

Sambucus mexicana
Blooms: March-June

These two species of elderberries are common in the coastal mountains. Coast red elderberry grows near the coast in moist shaded valleys. Blue elderberry is more common inland in mixed woodlands.

Both plants are characterized by the same distinct leaf pattern. The leaves are compound, with 6 to 8 finely toothed leaflets in an opposite pattern along the stem, ending in a final leaflet at the tip. In spring small, white, star-shaped flowers are borne in clusters 6 to 10 inches wide. Blue elderberry clusters are flat-topped and ripen into deep blue berries, while the red elderberry has dome-shaped clusters which ripen to scarlet red.

The light stems had several uses. Indians cut the plants back each fall so that in the spring the shoots would be straight enough to make arrows. Children hollowed out the pithy stems to make flutes and peashooters. Indians of southern California also used the elderberry stems to make a blackish dye for their basketry designs.

Although the blue berries are edible when cooked, most if not all of the red berries are poisonous.

The genus *sambucus* comes from the Greek word *Sambuke*, a musical instrument made from elder wood.

Coast Red Elderberry

Blue Elderberry

Blue Elderberry

CALIFORNIA BUCKEYE

Hippocastanaceae
(Buckeye) Family

Aesculus californica
Blooms: April-June

The habitat of California buckeye ranges from open dry slopes to wooded canyons. It is abundant throughout the Coast Range.

In late spring this is one of the showiest natives in this area. Fresh bright green leaves have replaced the bare limbs of winter. Characteristically, the leaves are palmate and composed of 5 to 7 leaflets. Large fragrant white flower spikes also begin to appear. By the end of summer, the leaves have fallen and the flowers are replaced by large brown chestnut-like fruits.

Inedible before leaching, the toxic nuts were used in the capture of fish by California Indians. Mashed nuts floated on the surface of the water acted to stupefy the fish, enabling them to be more easily caught.

GOLDEN CHINQUAPIN

Fagaceae
(Oak) Family

Castanopsis chrysophylla
Blooms: July-August

Golden chinquapin grows in dry woods and chaparral areas within the Coast Range.

The leaves of the chinquapin are tapered at both the base and tip, somewhat resembling California bay leaves. Like many chaparral plants, these leaves have a thick, leathery texture. A covering of small golden hairs on the underside of the leaf gives the golden chinquapin its name. The female flowers appear in late summer, are pollinated by male catkins, and produce clustered nuts in September. Each nut is surrounded with spiny bracts which give it the appearance of a large burr.

These sweet-tasting nuts can be roasted and eaten much like the chestnut. They are also a favorite of squirrels and other small animals who gather the nuts for their winter food supply.

California Buckeye

Golden Chinquapin

California Buckeye

CALIFORNIA HAZEL

Corylaceae
(Hazel) Family

Corylus cornuta ssp.*californica*
Blooms: (female) January-March

California hazel is a common shrub of shaded areas within the Coast Range. It can be found on wooded slopes as well as near streams and creeks.

Reaching 3 to 10 feet in height, these large shrubs have smooth slender stems and an open, spreading appearance. The alternate leaves have rounded bases and sharply pointed tips. Their softness and fine double-toothed margins are characteristic features.

In October and November, slender 1 to 3 inch long catkins appear; they release pollen from January to March. In late spring and summer the nuts mature, covered by a tubular bristly husk.

These nuts are not only edible, but are comparable in flavor to commercial filberts. They can also be ground into a meal and baked in bread.

Stems from this shrub were useful in basketry. To prepare the stems for weaving, the Indians either scraped them with a sharp rock or peeled the bark off with their teeth. After this procedure the lightweight stems were used as thread, rimhoops, or foundations for baskets, some as fine as flour sieves.

California Hazel

California Hazel

FLOWERS - PINK TO RED

WESTERN BURNING BUSH

Celastraceae
(Staff Tree) Family

Euonymus occidentalis
Blooms: April-July

Western burning bush is found near streams in the mixed evergreen and redwood forests.

An unusual leaf pattern easily distinguishes this large bush. The dark green leaves are arranged oppositely along the stem, with the last pair forming a distinctive fork. Brownish-red flowers, which ripen into deep red berries in late summer, hang on small stalks beneath the leaf junctions.

Another name for this shrub is "pawnbroker's bush," because the fruits hang down in clusters of three, resembling the pawn shop symbol.

CANYON GOOSEBERRY

Grossulariaceae
(Gooseberry) Family

Grossularia sanquineum
Blooms: March-June

Gooseberry grows in shady wooded areas, especially near creeks. Red spines cover the stems and berries of this shrub. Its palmate leaves are alternately arranged. Fuchsia-like in appearance, the small hanging flowers are red at the base with white petals.

All species in the genus have edible berries, high in Vitamin C, which were eaten raw or cooked in pies, preserves, and jellies. Also, like currants, the berries were dried, mixed with animal fat, and eaten while traveling.

Western Burning Bush

Western Burning Bush

Canyon Gooseberry

LOVELY CLARKIA

Onagraceae
(Evening Primrose) Family

Clarkia concinna
Blooms: April-August

LARGE GODETIA

Clarkia purpurea
Blooms: April-August

Clarkia is found on dry grassy or brushy slopes throughout the Coast Range. In early summer, these colorful wildflowers make attractive displays on the hillsides.

Growing along tall graceful stems, both lovely Clarkia and large Godetia have small alternate linear leaves. The narrow-based flower of lovely Clarkia has distinctly clawed rose to purple petals. Large Godetia a broader flower with large, rounded petals. They range in color from pinkish to purple-red, often having darker center spots. The flowers mature into leathery quadrangular capsules, releasing small brown seeds by late summer.

The numerous species of *Clarkia* were named after Captain William Clark of the Lewis and Clark expedition.

CRIMSON COLUMBINE

Ranunculaceae
(Buttercup) Family

Aquilegia formosa var.*truncata*
Blooms: April-June

This beautiful spring wildflower can be found on many moist wooded slopes in the coastal mountains. The nodding flowers often hang over trails and road banks, delighting many a passer-by.

Growing from a woody base, this perennial plant has variously lobed, toothed, basal leaves. Openly branching stems often reach 3 to 5 feet in height. The showy flowers hang singly from the stem, at the outermost tips of the branches. Extending between the spurred red petals are 5 bright orange-red sepals. Long yellow stamens hang down from the center.

Lovely Clarkia

Large Godetia

Crimson Columbine

WILD GINGER

Aristolochiaceae
(Birthwort) Family

Asarum caudatum
Blooms: March-July

Wild ginger grows in moist shady areas and can be found along many mountain creeks in the northern Coast Range.

Its distinctive heart-shaped leaves form a dense carpet, 4 to 6 inches high, on the forest floor. Hiding below these leaves is a small cup-shaped maroon flower formed by long calyx lobes.

Because of the aromatic stems and roots, ginger is widely used in cooking. Early settlers dried and grated the roots for a spice. Candy was made by boiling the roots in sugar. As a medicinal herb, it was considered a remedy for flatulence and whooping cough.

RED CLINTONIA

Liliaceae
(Lily) Family

Clintonia andrewsiana
Blooms: April-June

Red clintonia is found near streams and in other moist areas of the redwood and mixed evergreen forests.

Most of the year clintonia can be identified by its large glossy green basal leaves, which contrast sharply with the darker forest plants. In late spring, a large cluster of deep pink trumpet-shaped flowers tops a long naked stem, which often reaches 2 feet in height. These flowers ripen into dark blue berries from which the plant gets such common names as "bead lily" and "blueing balls."

Wild Ginger

Red Clintonia

Red Clintonia

CALIFORNIA BEE-PLANT

Scrophulariaceae
(Figwort) Family

Scrophularia californica
Blooms: February-July

The California bee plant is common in open places. Because of its weedy nature, it oftens inhabits disturbed soils such as road banks.

This plant often grows to heights of 3 to 5 feet, easily towering over the other vegetation. The toothed spear-shaped leaves grow oppositely along a square stem. Small, often inconspicuous, flowers appear in branched clusters in early spring. The unique flowers have fused petals, forming a small cup with the two upper lips extending outward. The reddish color of the petals and the presence of nectar at the base of each flower attract bees. It is also known by the common name "figwort."

SPOTTED CORAL ROOT

Orchidaceae
(Orchid) Family

Corallorhiza maculata
Blooms: April-July

STRIPED CORAL ROOT

Orchidaceae
(Orchid) Family

Corallorhiza striata
Blooms: April-July

Spotted coral root grows in extremely shady areas, while striped coral root can be found in the more open oak-madrone forests and on dry, brushy slopes. Both species are inconspicuous and seldom seen.

The coral roots are saprophytes which, unlike most plants, have no green parts for producing energy from sunlight, but instead receive nutrients from decaying material. Upon casual observation, both plants appear to be nothing more than short, slender, reddish-brown stalks, but closer scrutiny shows the small flowers which are the distinguishing feature between the two species. The petals of spotted coral root are covered with reddish-brown blotches, while striped coral root flower petals have distinct veins of the same color. When the root is exposed it resembles a marine coral, hence the common names.

California Bee Plant

Spotted Coral Root

Striped Coral Root

FLOWERING CURRANT

Ribes glutinosum

Blooms: March-April

Grossulariaceae
(Gooseberry) Family

Flowering or winter currant is usually found in shaded woods and along streams.

This shrub has alternate leaves and, unlike its relative the gooseberry, it has no spines on the stems. The pink flowers hang in clusters from stalks and mature to black berries in the summer.

Like many other berries, currants are edible either raw or cooked into jams. Many Indians dried these berries, mixed them with animal fat, and stored them for winter use.

The name currant comes from the Corinth region in England, where plants similar to our currants grow.

CALIFORNIA GROUND CONE

Boschniakia strobilacea

Blooms: May-July

Orobanchaceae
(Broomrape) Family

Ground cones are parasitic plants which get their nutrients from other plants' roots. This herb associates most often with the roots of madrones and manzanitas, inhabiting dry wooded slopes and chaparral.

Often inconspicuous, the 6 to 10 inch high ground cones are a dark reddish brown and blend with the surrounding duff. The leaves are scale-like, resembling the bracts of a pine cone, hence its common name. Small 2-lipped tubular flowers project between these scales.

Flowering Currant

Flowering Currant

California Ground Cone

CALIFORNIA FUCHSIA

Onagraceae
(Evening Primrose) Family

Zauschneria californica
Blooms: April-September

The California fuchsia is found growing on poor, rocky soil adjacent to the dry chaparral areas.

When not flowering, this perennial fuchsia is a nondescript gray-green plant which grows in clumps. Its small narrow leaves grow directly off erect stalks, having no leaf stems. Covered with fine white hairs which can vary dramatically in number and size, these leaves are well protected against excessive water loss. The California fuchsia is one of the few spectacular blooming plants of late summer and fall. Its trumpet-shaped flowers cover the tops of the erect stems with bright scarlet.

Like our domestic ornamental fuchsias, the California fuchsia produces nectar which is a major source of energy for hummingbirds. It is especially important since it is one of the few food sources available when they start their southern migration.

CALIFORNIA HEDGE NETTLE

Labiatae
(Mint) Family

Stachys bullata
Blooms: April-September

This native plant inhabits most wooded areas in the coastal mountains. It can grow so dense that plants can completely cover moist ravines and slopes.

The slender simple stem of the California hedge nettle has opposite leaves along its length. Covering both stems and leaves are small soft hairs which, unlike true nettle, don't sting. In spring, flowers are borne on whorls at the top of the stalks. These purplish-red blooms are 2-lipped, like most mints.

The leaves, either soaked or steeped in water, can be used in the treatment of wounds and sores.

California Fuschia

California Fuschia

California Hedge Nettle

INDIAN WARRIOR

Pedicularis densiflora

Scrophulariaceae
(Figwort) Family

Blooms: January-July

Common on the warm eastern slopes of the mountains, Indian warrior grows in dry oak woodland or chaparral.

The deep red flowers which call attention to this plant grow in dense spikes atop a 6 to 20 inch tall stem. Several of these flower spikes grow from the base of each plant. The fern-like, finely divided leaves are also mostly basal, although some smaller ones do grow on the erect stem.

According to legend, each of these beautiful plants grows for a fallen Indian warrior. The legend behind the genus name isn't quite so romantic. An old superstition that sheep became infested with lice when they ate this plant resulted in the name *Pedicularis*, meaning louse.

RED LARKSPUR

Delphinium nudicaule

Ranunculaceae
(Buttercup) Family

Blooms: April-June

Red larkspur grows in dry areas in the mixed evergreen forests. Although not commonly seen in all regions, occasional large concentrations of the bright red flowers can be found. The Toll Road Trail in Castle Rock State Park is an especially good place to see these plants.

When not in bloom, red larkspur is an inconspicuous plant with short, lobed leaves. However, in late spring several scarlet red flowers with long spurs appear at the end of a 2 to 3 foot long leafless stalk. A related species, coast larkspur, has brilliant blue flowers.

Another name for red larkspur is sleep root, since it was used as a narcotic to dull the sense of gambling opponents by California Indians.

Indian Warrior

Coast Larkspur

Red Larkspur

CALYPSO ORCHID

Orchidaceae
(Orchid) Family

Calypso bulbosa
Blooms: March-July

This beautiful plant grows in moist woods and bogs along the northern west coast. Although rare in the Santa Cruz Mountains, the orchid is very common in areas of Humboldt and Del Norte counties.

The Calypso orchid has a single leaf and a single flower. Growing from a fleshy root stock, the glossy-green basal leaf serves as an early indicator of where the flower will appear. In mid-spring, the pink flower grows atop a slender 4 to 6 inch long stem. The 3 sepals and 2 of the petals extend outward over a third petal. This lower petal is a large inflated sac-like lip which is mottled pink and white and has a ring of small hairs lining the inner edge.

The orchid gets its common name from the Greek goddess *Calypso*.

CHAPARRAL PEA

Fabaceae
(Pea) Family

Pickeringia montana
Blooms: May-August

As its common name implies, chaparral pea is a chaparral shrub, often forming dense thickets. Its colorful flowers brighten up ridgeline trails throughout the mountains.

This plant has small, tough, water-conserving leaves and stiff branches. Its sharp spines sometimes stab the unwary hiker. Growing singly along the stems, large showy pink flowers are pea-shaped. As the flower matures a seed pod forms, releasing black, oblong seeds. Chaparral pea easily spreads by underground stems and is well adapted to a fire-burned environment.

Calypso Orchid

Chaparral Pea

WOOLLY PAINT BRUSH

Scrophulariaceae
(Figwort) Family

Castilleja foliolosa
Blooms: March-August

Woolly paintbrush is common on rocky chaparral slopes.

Several erect unbranching stalks with linear, sometimes 3-lobed leaves grow from the woody base of each plant. Covering both stems and leaves, long white hairs form a dense woolly mat and protect against moisture loss. On the upper tips of the stalks are bright scarlet flowers which stand out against the somber chaparral background. Unlike most flowers, it is the sepals and leaf-like bracts below them which are brilliant red, hiding the smaller yellow-green petals.

Since Indian paintbrush is found in rocky chaparral areas which are favorite places for rattlesnakes, they were known by some Indian tribes as "snake's friend." Their bright flowers were thought to be the source of rattlesnake venom. Surprisingly, or perhaps not, they were also used as a love charm.

CALIFORNIA RHODODENDRON

Ericaceae
(Heath) Family

Rhododendron macrophyllum
Blooms: April-July

This showy native inhabits the dry slopes in its southern distribution, the Santa Cruz Mountains, but can be found growing in moist woods in the northwestern part of the State. It is common amid the redwoods of Humboldt and Del Norte counties.

The rhododendron is related to the more common western azalea. Often reaching heights of 10 to 12 feet, this evergreen shrub has leathery dark green leaves. Large fragrant clusters of bright pink flowers cover the shrub in late spring, maturing into cylindrical capsules. Brown-winged seeds are released in the fall.

Many varieties of the rhododendron are used commercially for garden ornamentals. The scientific name *rhododendron* is Greek and means "rose tree."

Woolly Paint Brush

California Rhododendron

SALMONBERRY

Rosaceae
(Rose) Family

Rubus spectabilis var.*franciscanus*
Blooms: March-June

Salmonberry is a common shrub along the margins of woods and streams from northern California to Alaska. It grows below 1,000 feet in elevation.

Growing from 3 to 9 feet in height, this shrub forms dense, impenetrable thickets. The stems have reddish-brown bark usually without prickles. The erect canes, however, are covered with short, straight prickles. The leaves, compound with 3 leaflets, have serrated edges and are 1/3 to 1/2 inches long. The salmon-colored flowers produce red or yellow berries in late summer. The berries are edible.

WOOD ROSE

Rosaceae
(Rose) Family

Rosa gymnocarpa
Blooms: April-September

Wood rose grows in shady areas of redwood and mixed evergreen forests.

A small shrub with slender stems and fine, straight prickles, wood rose grows 2 to 4 feet high. Its leaves, composed of 5 to 7 leaflets each, are arranged alternately along the stem. The dainty pink blossoms are about 1 inch wide. In autumn, these flowers are replaced with red urn-shaped rose hips.

Rose hips, which have 24 times as much vitamin C as oranges, were used by Indians to make a rose-colored tea for the relief of colds. Today rose hips are used commercially to make vitamin C tablets.

Some Indian tribes used tea from the leaves to relieve pains and colic. The petals can also be used for tea, as well as eaten raw or in jellies.

Another species of rose, the California rose (*Rosa californica)*, is also commonly found in these mountains and has the same uses as wood rose. In comparison with wood rose, the California rose has a fuller, brighter-colored flower, hairy leaf undersides, and curved prickles.

Salmonberry

Wood Rose

California Rose

REDWOOD SORREL

Oxalidaceae
(Oxalis) Family

Oxalis oregana
Blooms: February-September

Redwood sorrel is extremely abundant in the redwood and mixed evergreen forests, often covering the ground in thick carpets.

The clover-like leaves are extremely sun sensitive, folding down like umbrellas whenever it gets too hot. Solitary flowers grow on small stalks and turn from white to deep pink with age.

Both the stems and leaves can be eaten raw in salads or slightly fermented for a tangy dessert. Their sour taste is responsible for the Latin name *Oxalis*, or acid juice. The pioneers used these sour stems in a pie similar to rhubarb pie.

PACIFIC STARFLOWER

Primulaceae
(Primrose) Family

Trientalis latifolia
Blooms: March-June

The Pacific starflower can commonly be found within the redwood and mixed evergreen communities. Like redwood sorrel, the starflower often completely covers large areas of ground in these forests.

This plant is easily identified by its small whorl of leaves atop a slender stem, with 1 to 4 tiny pink star-shaped flowers growing from the center. Although somewhat resembling trillium, the starflower has thinner, smaller leaves, with 5 to 7 in a whorl.

Redwood Sorrel

Pacific Starflower

GIANT WAKE ROBIN

Convallariaceae
(Lily-of-the Valley) Family

Trillium chloropetalum
Blooms: February-April

The giant wake robin is usually found on brushy woody slopes. More common north of San Mateo county, it is seldom seen in the Big Basin Redwoods State Park area. One nice cluster grows next to the Castle Rock State Park automobile parking lot.

Although taller than western wake robin, this plant is similarly characterized by its whorl of 3 large dark green leaves which grow directly from the main stem. The flowers, composed of 3 erect petals, grow directly above these leaves and have no flower stalks. As they mature, the flowers darken to a deep red-purple.

While the deep maroon color of the flowers is common in the central area of these mountains, the more northern flowers are often tinged with green and are responsible for the genus name *Chloropetalum*, meaning "green petal." Like western wake robin, the underground stems can cause vomiting if eaten.

WINTERGREEN

Pyrolaceae
(Wintergreen) Family

Pyrola picta
Blooms: June-August

WINTERGREEN

Pyrola bracteata

This small herb usually grows on wooded slopes, often associating with Douglas firs and madrones.

Wintergreen, or red pyrola, grows from a slender root, sending up an often leafless stalk 6 to 7 inches in height. The delicate 5-part flowers have red to purple sepals and pink petals, which often have white margins. Giving the flowers a pendant appearance, 10 stamens grow around the long style.

Giant Wake Robin

Wintergreen (Pyrola picta)

Wintergreen (Pyrola brateata)

GNOME PLANT

Pyrolaceae
(Wintergreen) Family

Hemitomes congestum
Blooms: May-June

The gnome plant is a saprophyte that grows in the humus of redwood and mixed evergreen forests from Monterey county north to Washington state. Although uncommon, it is a pleasant find along the trail.

This fleshy herb grows as a dense cluster of stout scale-like leaves. Often the stems are underground or covered by surrounding duff. The leaves are white-pinkish in color and are topped by heads of tubular flowers. Surrounded by large bracts, the flower corolla can be 4 to 6-lobed with a prominent yellow stigma in the center.

As a saprophyte, the gnome plant derives its nutrients from decaying organic matter in the ground.

FIREWEED

Onagraceae
(Evening Primrose) Family

Epilobium angustifolium
Blooms: June-September

Fireweed grows along the entire Coast Range, becoming more common in the northern areas. Often growing in dense patches, it inhabits moist areas in mixed evergreen or redwood clearings as well as recently disturbed or burned areas.

This hardy perennial grows 2 to 5 feet tall, usually on upright, unbranched stems. Long, linear, "willowy" leaves densely cover the stem. On the upper portion of the plant are showy clusters of bright magenta, or occasionally white, flowers. Each flower has four sepals and four clawed petals. Maturing in the late summer, small pods release many tiny, parachuted seeds.

These minute seeds drift easily in air currents, enabling fireweed to be one of the first colonizers after burns. It was one of the first plants to cover the battle scars of Europe after World War II.

Gnome Plant

Fireweed

WESTERN BLEEDING HEART

Fumariaceae
(Fumitory) Family

Dicentra formosa
Blooms: March-July

This native grows in damp, shady areas along the coast of northern California to British Columbia. A nursery grown variety is often planted in gardens. Five species are native to California.

Arising from a fleshy rootstock, the western bleeding heart has long-petioled, finely-divided fern-like leaves. The clusters of rose-purple flowers are heart-shaped and hang from long, naked stems. Individual flowers are about 3/4 inch long with 4 petals, the two outer ones forming the outline of the "heart."

In gardens the bleeding heart grows rapidly and can be propagated easily by divisions of the root or from seed. This plant is also known as "Dutchman's Breeches."

HAIRY HONEYSUCKLE

Caprifoliaceae
(Honeysuckle) Family

Lonicera hispidula
Blooms: May-June

Honeysuckle is found along streams and in moist or shaded wooded areas below 2,500 feet elevation.

This climbing woody vine has reddish stems which can grow up to 20 feet long when supported by other vegetation. Bare along much of their length, these stems support 2 1/2 to 4 inches long leaves near their tips. The uppermost leaves are fused and the remainder have very short petioles. Arranged in many whorls, the purplish-pink trumpet-shaped flowers produce translucent red berries in late summer.

This honeysuckle is closely related to the ornamental honeysuckle, which is native to Japan.

Western
Bleeding Heart

Hairy Honeysuckle

Hairy
Honeysuckle

RED THISTLE

Compositae
(Composite) Family

Cirsium proteanum
Blooms: May-July

The red thistle, one of more than 200 species of thistles growing in northern hemisphere, is native to California. It grows on dry slopes below 10,000 feet.

This thistle grows from a stout, branched stem, usually 1-1/2 to 3 feet tall. The lance-shaped leaves are prickly and covered with fine hairs. Solitary flower heads are composed of short red spines surrounded by prickly bracts. They are often laced with cobwebby hairs.

OWL'S CLOVER

Scrophulariaceae
(Figwort) Family

Orthocarpus purpurascens
Blooms: March-May

Often coloring entire meadows with a deep purple haze, owl's clover grows in grassland and open woodlands from central California north to Mendocino county.

The pale to deep magenta flowers grow atop short unbranched stems in dense tufts, like fat paintbrushes. Resembling tiny perched owls, each flower has an upper hooked lip and a lower puffy lip. The lower lip is 3-lobed with 3 small teeth. Its tip is colored yellow or white. Hiding beneath the showy flowers are small leaves, each divided into several small segments. The 1/2 inch long seed capsule gives this genus its Greek name which means "straight fruit."

Owl's clover is bee pollinated. Its delicate purple coloring is due to anthocyanin, a pigment which unfortunately fades if picked.

Red Thistle

Owl's Clover

COMMON LINANTHUS

Polemoniaceae
(Phlox) Family

Linanthus androsaceus
Blooms: April-July

Common linanthus grows well in the lean soil of open grasslands and chaparral from Monterey county through northern California.

Often found in clusters, this small, somewhat spindly annual grows from 2 to 12 inches in height. Its slender stems are usually simple, but occasionally branch. Ranging from deep pink or lilac to white or yellow, the 5-lobed flowers are 1/2 to 3/4 inch long. From above, they appear round and flat, but a side view reveals the long slender flower tube. A large cluster of leaves rings the stem directly below the flower. Other leaves grow in scattered pairs along the stem. Each is finely divided, with 5 to 9 narrow lobes.

With bright flowers and delicate leaves, members of the Phlox family are often grown as ornamentals.

FIRE-CRACKER FLOWER

Amaryllidaceae
(Amaryllis) Family

Brodiaea ida-maia
Blooms: May-July

The fire-cracker flower grows in open grassy areas in redwood and mixed evergreen forests in the northwestern portion of the state. It generally can be found between 1,000 and 4,000 feet elevation.

Growing from underground corms, this perennial plant has long thin linear leaves ranging from 4 to 15 inches in length. A separate flowering stalk emerges in the spring from which hangs a cluster of bright red tubular flowers. The red perianth has greenish-white segments at the tip, which curve backward as the flower matures.

The *Brodiaeas* make up a large genus in western North America. As well as being a beautiful sight on a spring day, many species were utilized by California Indians for food. The roasted corms (or bulbs) and seeds are edible from several species.

Common Linanthus

Fire-Cracker
Flower

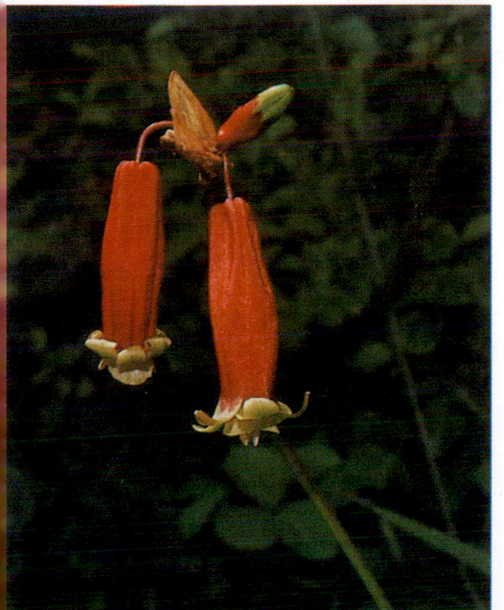

REDWOOD PENSTEMON

Scrophulariaceae
(Figwort) Family

Penstemon corymbosus
Blooms: June-October

This small shrub grows on rocky slopes and cliffs along the Coast Range from Monterey to Del Norte counties as well as in the Sierra foothills below 5,000 feet.

The redwood penstemon has small, dark green, leather-like leaves growing oppositely along the stem. The leaf margins are coarsely serrated. Growing from the end of the stem are clusters of reddish tubular flowers. The flowers are five-lobed, two form the upper lip, and three form the spreading lower lip. The five stamens are hairy at the base; four are fertile and one is sterile and tipped with yellow hairs.

The genus name, *penstemon,* is Greek and refers to the presence of five stamens. Like most red tubular flowers in nature, the penstemon provides an important food source for hummingbirds with its nutrient-rich nectar.

CALIFORNIA MILKWORT

Polygalaceae
(Milkwort) Family

Polygala californica
Blooms: April-July

The California milkwort is a common herbaceous plant of the redwood and evergreen forests. It is often seen hanging over trail ledges as well as on the forest floor. It is most notable when blooming during the early summer months.

This is a small plant, reaching 6 to 8 inches in height. The leaves, simple in design, are arranged alternately upon the stem and are 1/2 to 1 1/2 inches long.

The terminal clusters of rose-purple flowers are pea-like; the 5 sepals and petals are united at the base, then form a broad lower keel and side wings. The superior ovary yields an elliptical and notched seed capsule that splits open along the margin, releasing the small seeds. The name, *Polygala,* is from ancient Greece. Polus means "much" and gala "milk," which refers to a shrub thought to stimulate milk flow.

Redwood Penstemon

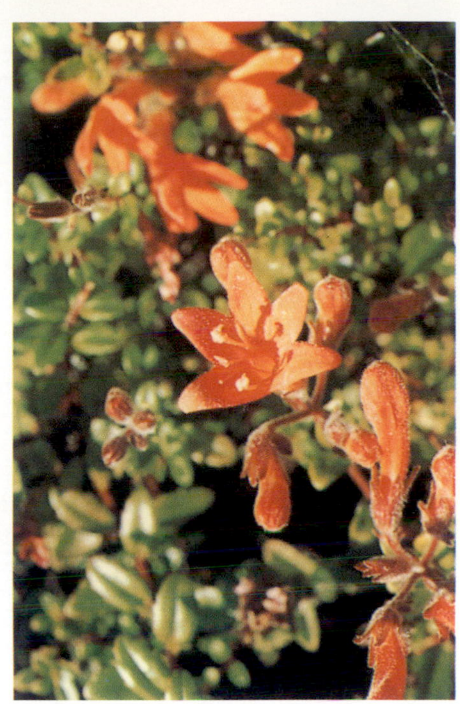

Redwood Penstemon

California
Milkwort

FLOWERS - BLUE TO PURPLE

BLUE-EYED GRASS

Iridaceae
(Iris) Family

Sisyrinchium bellum
Blooms: February-May

Blue-eyed grass is common in open meadow areas, often associated with oak woodlands.

This plant is a classic demonstration of the problems with common names. Blue-eyed grass is not a grass at all, but an iris. Rather than having blue "eyes," the flowers are dark purple with yellow "eyes" in the center. These flowers are perched singly on short stalks off the main stem. Long, slender leaves that clasp the stem are often hidden by surrounding grasses. The albino variety in the photograph is quite uncommon, but it is striking when growing alongside the blue.

Because pigs sometimes grub at the woody roots, the genus name *Sisyrinchium* means pig snout; the species name *bellum* means handsome.

Tea made from blue-eyed grass was used as an early remedy for fever reduction.

BLUE DICKS

Amaryllidaceae
(Amaryllis) Family

Dicheleostema pulchellum
Blooms: February-May

Blue dicks are found in open grassy areas. They are one of the most common of the spring wildflowers.

Conspicuous, tightly-clustered flowers grow on a naked stem about a foot tall. Much less noticeable are the long slender leaves, which grow from the base of the plant.

The perennial bulbs were known as Indian onions or potatoes. They were dug out with a digging stick, then eaten raw, boiled, or roasted. The small black seeds were roasted like popcorn in a mix called pinole, which is similar to our present day trail mix.

Blue-eyed Grass Kathy Lyons

Blue Dicks

Blue Dicks

BLUE WITCH

Solanaceae
(Nightshade) Family

Solanum umbelliferum
Blooms: January-September

Blue witch is found on dry rocky slopes in chaparral or in open areas of oak woodlands.

This woody shrub has hairy leaves which are grayish green. The pinwheel-shaped flowers are deep purple with bright yellow anthers. On the base of the petals surrounding the anthers are pairs of small white dots with green spots in their centers.

Although a close relative of the tomato and potato, this plant is poisonous if eaten. The entire plant is high in solanine, an alkaloid which is also present in the leaves and stems of tomatoes and potatoes. Solanine poisoning can cause minor symptoms such as drowsiness, trembling, weakness, nausea, and abdominal pain, or it can lead to serious problems such as paralysis, unconsciousness, or death.

CALIFORNIA HAREBELL

Campanulaceae
(Bellflower) Family

Asyneuma prenanthoides
Blooms: June-September

This perennial grows on dry wooded slopes in redwood and mixed evergreen forests. It is a common summer wildflower.

The California harebell has simple, toothed, lance-shaped leaves growing alternately along a slender stem. Bright blue tubular flowers hang in scattered clusters off the stem. A long central style extends beyond the blue petals, giving it a bell-like appearance.

The family name, *Campanulaceae*, comes from the Latin word *campana*, meaning little bell. Many species of this family are used as ornamentals.

Blue Witch

California Harebell

WESTERN HOUND'S TONGUE

Boraginaceae
(Borage) Family

Cynoglossum grande
Blooms: February-April

Hound's tongue is one of the first flowers of spring and is common in wooded mixed evergreen forests throughout these mountains.

This native has large, simple leaves arising mostly from the base of the plant. Small hairs on the surface give the leaf a rough texture. The shape and texture somewhat resemble a dog's tongue, hence the common name. Clusters of small, bluish-purple flowers are borne atop a long central stalk and mature into spiny nutlets. The tiny hooks cling easily to animal fur, aiding in seed dispersal.

CALIFORNIA FETID ADDER'S TONGUE

Convallariaceae
(Lily-of-the-Valley) Family

Scoliopus bigelovii
Blooms: February-March

Fetid adder's tongue grows in moist redwood valleys, especially along streams.

Since this small plant is one of the earliest to flower in the spring, few hikers see the exquisite blooms. Composed of sepals which are stamped with distinct, dark purple veins and erect, dark purple, horn-like petals, the blooms grow on short stalks between the basal leaves. These large oval leaves are pointed at the tip and have purple blotches scattered over the glossy green surface. Individual plants usually bear only 2 leaves, but occasionally have 3.

Although the sight of the flower is a treat, its smell, which probably attracts flies as pollinators, is not so pleasant and is responsible for the name. Another common name, slink pod, refers to the fact that the slender stems droop to the ground as the flower's pods mature. The genus name, *Scoliopus*, meaning crooked foot, also refers to this drooping stem.

Western Hound's Tongue

California Fetid Adder's Tongue

LUPINE

Lupinus ssp.
Blooms: March-August

Fabaceae
(Pea) Family

Several species of lupine are found in this area. Although probably most common in open meadows, these plants can be found in almost every habitat from sandy beaches to shady forests. The various species range from being quite common to extremely rare.

Although their growth forms vary from small annual or perennial herbs to large shrubs, lupines have several distinguishable features in common. The leaves, which are always alternate and compound palmate, are the most easily identifiable characteristic of these plants. Also, all lupines have flowers arranged in whorls near the top of erect stems. Individual flowers are pea-like with 5 parts. Reflexed back behind the other petals, the large upper petal is called the banner. In front of this petal are 2 smaller petals called wings which surround 2 even smaller, joined petals called the keel. Most of our local lupines are various shades of lavender to blue, although a few are shades of yellow or white.

The name *Lupinus* comes from *lupus*, meaning wolf, because these plants were once thought to destroy the soil. However, this couldn't be further from the truth, since lupines help the soil both by stabilizing it with their deep roots and by building up its nitrogen supply with the bacteria in its root nodules.

Although the seeds reportedly were boiled and used to treat urinary disorders, they often contain dangerous alkaloids.

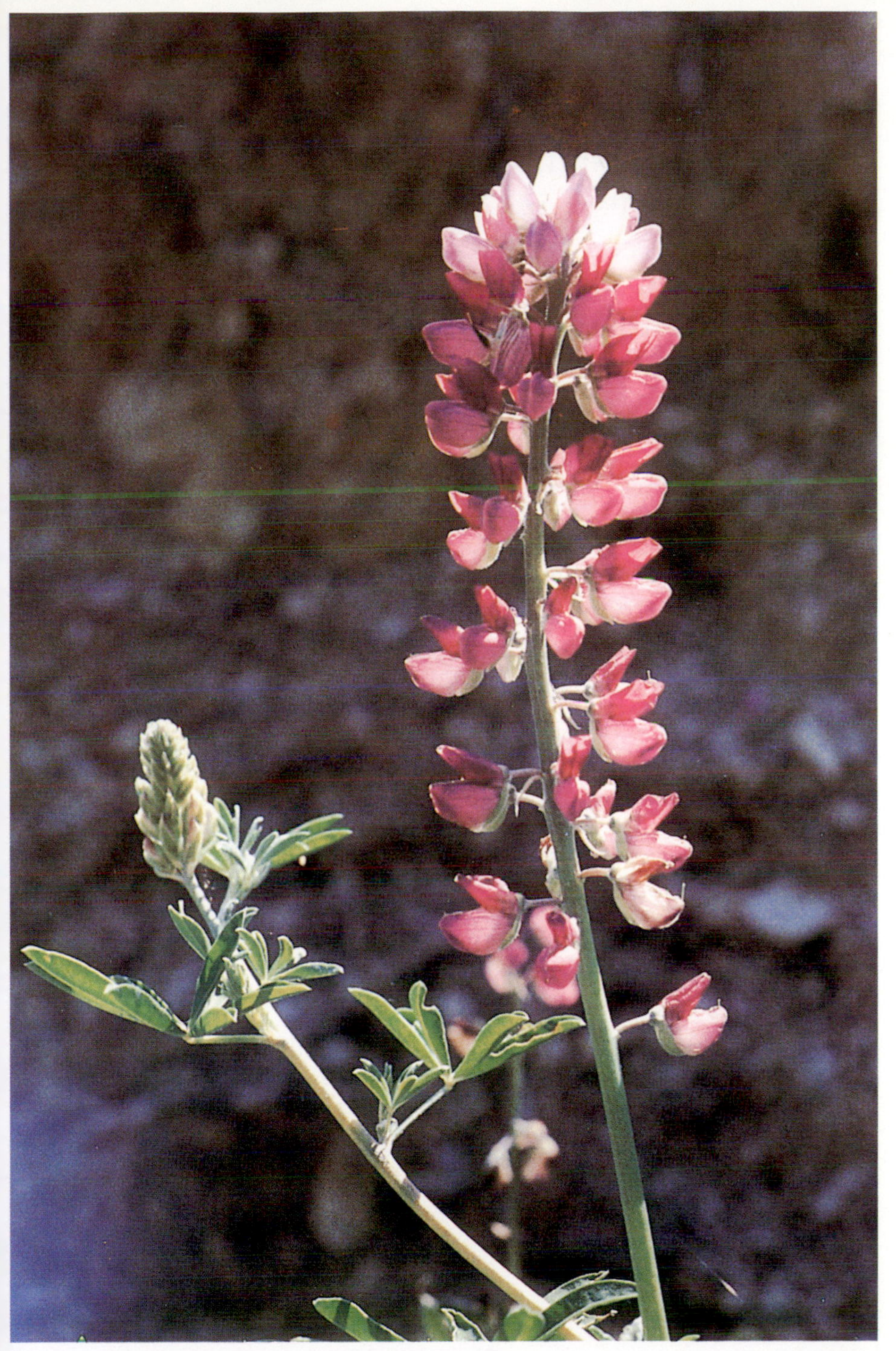

Lupine

CALIFORNIA WILD LILAC

Rhamnaceae
(Buckthorn) Family

Ceanothus thyrsiflorus
Blooms: March-April

WARTY-LEAVED CEANOTHUS

Rhamnaceae
(Buckthorn) Family

Ceanothus papillosus
Blooms: March-April

Lilac is common in drier regions of several communities. It can be found in chaparral, mixed evergreen forests, meadows, oak woodlands, and redwood forests.

Several species of lilac grow in the Coast Range, including both the white and the bluish-purple varieties. All species are large, woody shrubs with fragrant flower clusters composed of tiny individual flowers.

Blue-blossom ceanothus (*Ceanothus thyrsiflorus)* is common in more moist, shaded regions of the mountains, sometimes forming almost pure forests with shrubs 20 feet high. Its oval-shaped leaves are bright green and have 3 major veins converging at the base.

Warty-leaved ceanothus (*Ceanothus papillosus)* is common in chaparral and along forest margins. Its deep purple flowers and small warty leaves make it the most easily identifiable of the lilacs.

Early Indians had several uses for these shrubs. Flowers and leaves were made into tea, while the bark was used to make a tonic. Leaves were used as tobacco and roots made a red dye. Several tribes used the stems as foundations for their baskets.

When the blossoms are mixed with water and rubbed vigorously they make a fragrant soap. In some Indian tribes, this soap was used by the bride and groom to wash each other's hair as part of the wedding ceremony.

Warty-leaved Ceanothus

California Wild Lilac

HENDERSON'S SHOOTING STAR

Primulaceae
(Primrose) Family

Dodecatheon hendersonii
Blooms: February-May

Shooting stars grow on open slopes and in woods in the coastal mountains. Especially large concentrations can be found blooming near small intermittent springs and creeks.

This plant is well known for its attractive and colorful flower clusters. Hanging atop slender stems, they have maroon to purple petals which are bent backwards to an exposed dark central stamen. Rings of yellow and black encircle the base of the petals. The glossy green oval leaves have long stems and arise in a cluster around the base.

Either boiled or roasted, the roots and leaves are edible.

CALIFORNIA TOOTHWORT

Cruciferae
(Mustard) Family

Dentaria californica var.*californica*
Blooms: December-May

Toothwort is a common inhabitant of all shaded areas within the coastal mountains. Although extremely variable, it is generally one of the earliest spring wildflowers.

This slender plant grows from a fleshy underground stem and has two characteristic leaf patterns. Broad oval leaves arise independently from the root base, while slender lance-shaped leaves grow from the erect, unbranched flower stalk. These generally consist of 3 leaflets. White to lavender 4-part flowers grow in clusters from the top of the stem, maturing into long slim seed pods. Another variety of *Dentaria*, milkmaids (var.*integrifolia)*, is also found in this area, and the two freely hybridize.

The rootstocks are edible raw and are often added to salads. The name *Dentaria*, meaning tooth, pertains to these odd-shaped, toothed, underground stems.

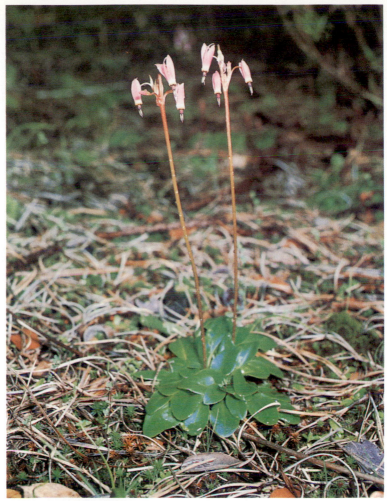

Henderson's Shooting Star

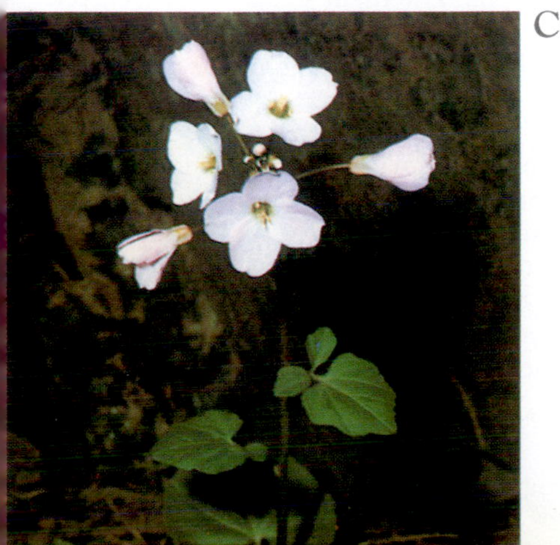

California Toothwort

VETCH

Fabaceae
(Pea) Family

Vicia ssp.
Blooms: March-July

Many species of vetch are found in these mountains, most of them natives of Europe. American vetch (*Vicia americana* var.*oregona)* is the most common species. Vetch usually grows along trails and roadsides or in other disturbed areas.

The leaves of this trailing vine are composed of alternate linear leaflets and usually have climbing tendrils. Similar to the garden pea, to which they are related, the flowers of most of our species range from pale lilac to deep red-purple, although a few species have at least some white. The flowers are either arranged in rows on a long stem or grow singly on short stems from the leaf axis.

The genus name *Vicia* means to bind together and refers to the way in which the tendrils wrap around other plants.

Young seeds and shoots can be cooked and eaten like domestic peas.

YERBA SANTA

Hydrophyllaceae
(Waterleaf) Family

Eriodictyon californicum
Blooms: April-July

Yerba santa is one of the most abundant plants of the dry chaparral regions. However, it is an opportunist and can occasionally be seen growing along shaded streamsides.

A tall shrub, sometimes reaching 5 feet in height, yerba santa has numerous erect stalks growing from a small short trunk. The main stems are often black from a covering of sooty fungus. Leaves are lance-shaped, thick and sticky. The tubular flowers are soft lavender and grow in clusters from the tops of the stems.

Yerba santa, meaning holy weed in Spanish, was named by missionaries when they were told by Indians of its many medicinal uses. A bitter tea made from the leaves was used to treat everything from tuberculosis to rheumatism, including coughs, sore throats, and asthma. A weaker tea was used as a blood purifier. Fresh leaves in a poultice were bound on sores, and a strong solution brewed from the leaves was used to soothe sore and tired limbs. Some tribes smoked or chewed these leaves like tobacco.

Vetch

Yerba Santa

WOOLLY BLUE CURLS

Labiatae
(Mint) Family

Trichostema lanceolatum
Blooms: June-October

This herbaceous plant can be found growing on dry slopes, open fields, and on overgrown roadways from northwest Oregon south to Baja California. Woolly blue curls also bears the names "camphor weed" and "vinegar weed," because of its pungent, far-reaching odor.

Woolly blue curls is a tall, leafy plant, often reaching heights of 1 to 3 feet. Narrow leaves, 3/4 to 2 1/2 inches long and covered with minute hairs, are arranged oppositely along the stem and point upward. Clusters of 2 to 3 flowers arise from the leaf axils. The pale blue flowers, about 1/2 inch long, are tubular with the slender floral tube curving up abruptly into an arc. The 4 stamens, when mature, project through the upper floral lobes and gracefully curve backwards.

A poultice was made by California Indians from the leaves and flowers of woolly blue curls to treat colds. Leaves could also be used for an aching tooth by packing them around the inflammation. As an aid in catching fish, crushed leaves thrown into a stream would stupefy fish, enabling them to be easily caught. This method, however, is unlawful now because of its adverse effect on the stream's ecology. The pollen from this plant is an important food source for honeybees.

Woolly Blue Curls

FLOWERS - WHITE TO CREAM

BEDSTRAW

Rubiaceae
(Madder) Family

Galium ssp.
Blooms: March-April

Bedstraw grows in moist, shady areas of the redwood and mixed evergreen forests.

It is a spreading vine characterized by square stems and small linear leaves, which grow in whorls around the stems. Because of its small, backward hooks this vine is an excellent climber and often grows over other plants. The tiny, whitish-green flowers grow in clusters of 2 to 5 on a short stalk from the leaf whorl axis.

In some countries this sweet-smelling plant is still used for stuffing mattress, hence the name bedstraw. One species is even credited with having lined Christ's manger in Bethlehem.

Bedstraw has many other uses. Along with rennet, it was once used in the preparation of cheese. The seeds of some species are used as a coffee substitute when roasted and ground. The roots can be used to make a purple dye.

COFFEE BERRY

Rhamnaceae
(Buckthorn) Family

Rhamnus californica
Blooms: May-July

The coffee berry is a hardy shrub which prefers dry wooded areas or chaparral. It is commonly seen along many mountain trails.

This shrub has alternate leaves 2 to 5 inches long, and like most dry-region plants the leaves are hard and leathery to prevent moisture loss. In fall, clusters of greenish flowers ripen into black berries.

This shrub is a close relative of the cascara shrub which is cultivated commercially and used in the preparation of laxatives. The berries are edible either raw or cooked and are quite nutritious.

Bedstraw

Coffee Berry

WESTERN AZALEA

Ericaceae
(Heath) Family

Rhododendron occidentale
Blooms: June-September

This showy native forms dense thickets along streams and in moist meadows in redwood and mixed evergreen forests.

In spring this large deciduous shrub produces bright green leaves. Unlike its relative, the California rhododendron, the leaves are thin and somewhat glandular.

In locations with ample sunlight, large showy clusters of creamy-white or pinkish-white flowers cover the bush. The strong fragrance from these flowers is a treat for hikers along the trails. All parts of this plant are poisonous.

COMMON BUCK BRUSH

Rhamnaceae
(Buckthorn) Family

Ceanothus cuneatus var.*dubius*
Blooms: February-August

As it name implies, common buck brush, one of several white varieties of lilac, is often found in these mountains. It is abundant in chaparral, sometimes forming impenetrable thickets.

This large woody shrub has dark green, smooth-margined leaves which are thick and leathery to minimize water loss. The leaves grow on spine-like branchlets which often spear unwary hikers. The small white flowers grow in fragrant clusters. Uses of this shrub are described in the Blue-Flowering section under California wild lilac.

Western Azalea

Common Buck Brush

ELK CLOVER

Araliaceae
(Ginseng) Family

Aralia californica
Blooms: July-August

Elk clover, or spikenard, is usually found along streamsides or in very moist redwood or mixed evergreen forests.

This spreading shrub can reach up to 10 feet tall or more. Its alternate leaves have stems as much as 1 foot long. The leaves are compound, consisting of 3 separate groups of leaflets with 3 to 5 leaflets in a group. Individual leaflets can sometimes be as large as 5 inches wide by 7 inches long. In early summer, loose clusters of small white flowers bloom on the tips of the branches. Later these mature into purplish black berries.

CALIFORNIA BLACKBERRY

Rosaceae
(Rose) Family

Rubus ursinus
Blooms: March-August

Blackberries are usually found in sunny areas near water in most of the plant communities. They also form sprawling thickets in open meadows.

This thorny vine has stout, erect stems entwined with thinner trailing shoots. Most of the leaves are compound with 3 pointed leaflets. However, smaller simple leaves may also be found on flowering stems. Although sometimes confused with poison oak by beginning botanists, blackberry leaves and stems have thin prickles, while poison oak does not. The white, rose-like flowers produce juicy black berries in late summer.

The most common use of blackberries is as a food, either eaten raw or cooked in pies, jellies, or preserves. Also, young shoots can be sliced and eaten in salads. Blackberry brandy was used by early settlers as an efficient cure for diarrhea. A less familiar use was in the making of a black dye for basketry by the Luiseno Indians (a West Coast tribe).

Elk Clover

California Blackberry

California Blackberry

CHAMISE

Adenostoma fasciculatum
Blooms: May-June

Rosaceae
(Rose) Family

Chamise is one of the most common shrubs in the dry rocky chaparral communities.

Well adapted to the dry sunny habitats in which it grows, chamise has small, thick, leathery leaves. The highly branched stems are capable of regeneration by stump sprouting after a fire. In addition, this shrub exhibits an interesting survival mechanism called alleopathy, which is characteristic of many chaparral plants. Growth-inhibiting toxins are produced in the leaves, then washed into surrounding soil during rains to eliminate close competition.

In early summer, tiny white flowers grow profusely in panicles on the upper stems. By August, the dried flowers turn a beautiful rusty orange.

Although this tough plant isn't much of a delicacy, young stems can be tenderized by boiling for several hours, then seasoned, and eaten like a vegetable.

CHINESE HOUSES

Collinsia heterophylla
Blooms: April-July

Scrophulariaceae
(Figwort) Family

Chinese houses inhabit open or shaded slopes throughout the coastal mountains. They are often associated with oak woodlands and grassy knolls.

This annual grows from 1/2 to 1 foot in height. Slightly toothed, oblong leaves grow opposite on the slender stem. Blooming in late spring, clusters of 2 to 5 flowers grow in whorls. The individual flower parts are fused, forming 2 major lips. The upper lip is usually whitish and the lower lip is a deep lavender. The beautiful shape of the flowers gives the plant its common name, for they resemble Chinese pagodas.

Chamise

Chinese Houses

POISON HEMLOCK

Umbelliferae
(Parsley) Family

Conium maculatum
Blooms: April-August

Poison hemlock needs a sunny, open area to grow and therefore is often found in areas which have been disturbed by humans.

This tall plant has parsley-like leaves with a smooth, purple-splotchy stem. These stems can grow 10 to 12 feet high, terminating in large, white, umbel-shaped flower heads.

This native of Europe is extremely poisonous, producing instant paralysis resulting in death. Since hemlock somewhat resembles the edible cow parsnip and sweet fennel, novice gatherers should beware. Legend has it that poison hemlock was used to poison Socrates and to this day its purple-blotched stem is branded by his blood.

INSIDE-OUT FLOWER

Berberidaceae
(Barberry) Family

Vancouveria planipetala
Blooms: April-September

The inside-out flower inhabits moist areas, often associating with year-round streams and creeks.

These small delicate white flowers are rather inconspicuous. Growing from thin branching stems, they have reflexed sepals and petals that expose the central stamens, giving the blooms the appearance of being inside-out. The finely-toothed leaves grow off thin black stems, resembling those of maidenhair fern.

Poison Hemlock

Inside-out Flower

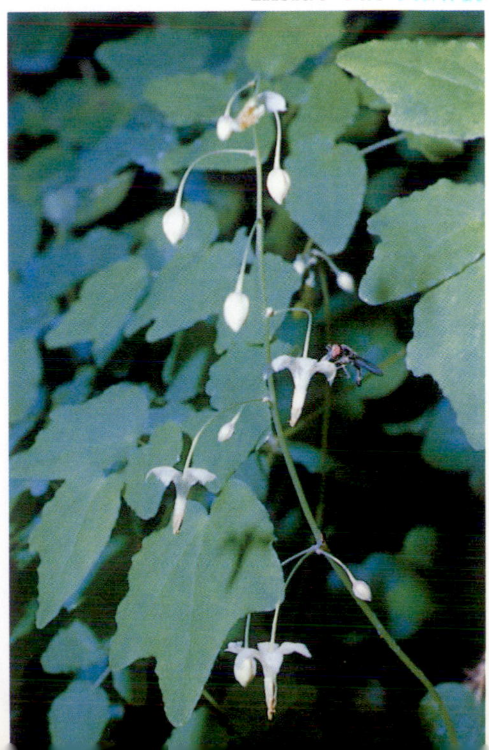

GLOBE LILY

Liliaceae
(Lily) Family

Calochortus albus
Blooms: April-May

Globe lilies grow in shaded dry areas under oak or mixed evergreen forests. Since they are somewhat uncommon and inconspicuous except when blooming, these small plants are seldom seen.

Globe lilies are bulb plants distinguished by a long, single, linear leaf growing from the base, with smaller leaves growing along the stem. The slender stems have nodding white flowers which branch off on short stalks. They are also knows as fairy lanterns because of their resemblance to lantern globes.

The perennial bulbs are nutritious and were eaten by California Indians. Although edible raw, they were most often boiled, roasted, or steamed in fire pits. Dried, they could be ground into flour for later use. Because of the globe lilies' limited distribution and the numerous other food sources available, these beautiful flowers should be left for all others to enjoy.

HOOKER'S FAIRY BELLS

Convallariaceae
(Lily-of-the-Valley) Family

Disporum hookeri
Blooms: March-May

Hooker's fairy bells have a wide range of habitats. They can be found in shady areas of redwood and mixed evergreen forests and also in oak woodlands.

This low-spreading plant is sometimes confused with Solomon's seal. However, it can be distinguished by its branching form. Small greenish-white flowers hang in pairs from the terminal leaves like hidden bells. In late summer, the flowers ripen to bright scarlet berries.

The two seeds in each cavity of its ovary give Hooker's fairy bells their genus name *Disporum*, while their species name is in honor of W.J. Hooker, a famous English botanist.

Globe Lily

Hooker's Fairy Bell

Hooker's Fairy Bell

WESTERN COLTSFOOT

Compositae
(Sunflower) Family

Petasites palmatus
Blooms: March-May

Coltsfoot is common in redwood forests near the edges of streams. It is especially abundant along Little Butano Creek in Butano State Park and along the upper Waddell Creek in Big Basin Redwoods State Park.

Deeply lobed, palmate leaves grow at the end of a 6-inch long stem which rises directly from the perennial rootstock. The leaf undersides are covered with soft white hairs. Blossoming at the end of a long stem with leaf-like bracts, the white flower appears somewhat like tightly-clustered dandelion seed heads.

Indians living away from the ocean had an interesting use for the plant. They would place the wilted leaves on a redwood burl over hot coals. In the slow heat of the fire the leaves burned down to ashes of almost pure salt.

MINER'S LETTUCE

Portulacaceae
(Purselane) Family

Montia perfoliata
Blooms: February-May

Miner's lettuce is fairly common in moist wooded areas.

This plant is easily identifiable by its united, saucer-like leaves which grow on 3 to 8 inch long stems. Tiny white flowers bloom on stalks from the leaf centers.

As the name implies, miner's lettuce can be eaten raw in salads or boiled like spinach. A common practice of California Indians was to place the plant near red ant hills. As the ants crawled over the leaves, they left behind a vinegary flavor like a salad dressing.

The Indians also made a tea from the leaves, which was used as a laxative.

Western Coltsfoot

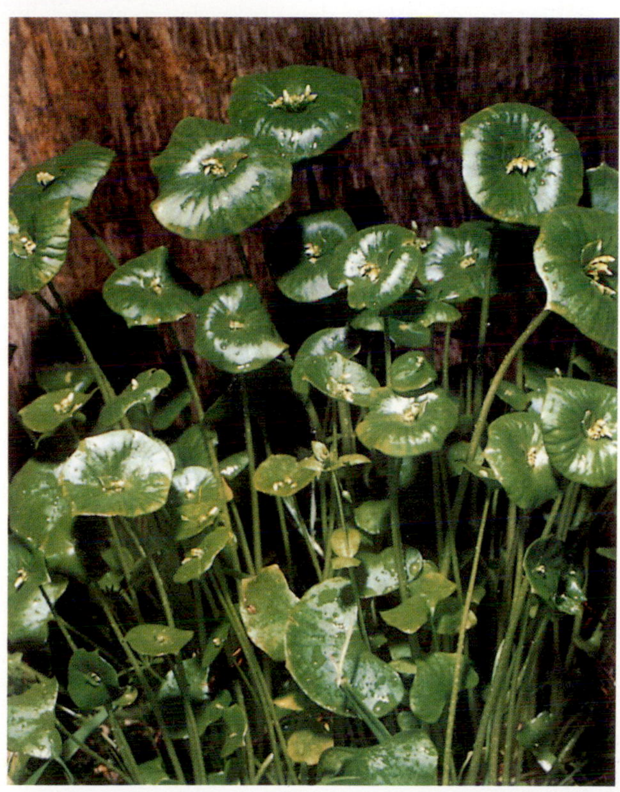

Miner's Lettuce

HUCKLEBERRY

Vacciniaceae
(Huckleberry) Family

Vaccinium ovatum
Blooms: February-June

Although most often found in redwood and mixed evergreen forests, huckleberry is extremely adaptable and can also be seen in chaparral. It grows profusely throughout the entire Coast Range.

This woody evergreen shrub forms large thickets which cover the forest floor. The alternate dark green leaves are thick and leathery enough to survive chaparral conditions. In spring, small white bell-shaped flowers hang below the branches, then ripen to deep blue berries in autumn. With the extra sunlight they receive, the chaparral area plants tend to have much sweeter berries.

Except for 10 small, hard seeds, the fruit is much like the blueberry and has the same uses. The berries are excellent raw, or cooked in pies, preserves, and jellies. In some places they are grown commercially. As the story goes, Mark Twain so relished huckleberry preserves that he named one of his most famous characters, Huckleberry Finn, after the plant.

RED HUCKLEBERRY

Vacciniaceae
(Huckleberry) Family

Vaccinium parvifolium
Blooms: March-June

The red huckleberry or billberry grows in the northern portion of the redwood region. It grows as a 3 to 10 foot high deciduous shrub. The branches, slender with sharply angled branchlets, have thin and pale green leaves arranged alternately along them.

Greenish-white flowers usually hang in pairs from the slender stems, maturing into flavorful bright red berries in late summer.

Huckleberry

Red Huckleberry

MOUNTAIN IRIS

Iridaceae
(Iris) Family

Iris douglasiana
Blooms: April-June

Mountain iris commonly grows in mixed evergreen forest, grassland, and open woodland.

The showy flowers range from blue to cream-white and are situated atop tall, fibrous stalks, sometimes with side branches. As is typical with all irises, the leaves clasp to the stem, which can be from 12 to 24 inches in height. Many similar species naturally hybridize, which can make identification difficult.

The fibers from the side ribs of the leaves were often twisted together by Indians to make a surprisingly strong string or rope. For small-game hunting, a bola was made by hanging small bones from the ends of the string. This was then swung toward the animal, entangling it and bringing it down. The tan-colored string was also woven together to form fishing nets and game snares. The northern Indians used young leaves as a wrap for their babies.

It should also be noted that all parts of the iris are poisonous.

STINGING NETTLE

Urticaceae
(Nettle) Family

Urtica holsericea
Blooms: May-October

Nettles grow along streams and in moist redwood forests.

These plants have large, spear-shaped opposite leaves which are coarsely toothed. They grow to almost 6 feet in height and are covered with small, poison-filled hairs which inflict extremely painful stings. White flower clusters hang from the junction of leaves and stems.

Despite the sharp poison hairs, nettle leaves and stems are edible when steamed or boiled and make a good spinach substitute. A mixture of salt and a strong solution of nettle was used to make rennet.

Mountain Iris

Mountain Iris

Stinging Nettle

MANZANITA

Ericaceae
(Heath) Family

Arctostaphylos ssp.
Blooms: various species bloom from November to May

Manzanita is one of the most common shrubs in chaparral. It grows on almost every dry ridgeline in the mountains.

These shrubs seem to be undergoing rapid evolution which results in many species, some of which are very similar and some very different. This makes it difficult to distinguish between species. However, all are characterized by reddish bark, with an outer skin which periodically peels off. Thick, leathery leaves are arranged alternately on the woody stems. Small clusters of bell-shaped flowers hang from the terminal tips of the branches.

Manzanita means little apple in Spanish and was so named because of its berries. Like several other wild berries, these fruits are edible and can be eaten raw or cooked in preserves, pies, and stews. When scalded, crushed, and strained they make a spicy cider, and can also be made into wine.

One species of manzanita has leaves and bark which were used as tobacco by West Coast Indians. Its leaves were also used for treatment of urinary tract disorders and are still used today as an ingredient in astringents. Because of the high tannic acid content of the leaves, they were once used in the tanning industry and are still so used in Russia.

WESTERN MORNING GLORY

Convolvulaceae
(Morning Glory) Family

Convolvulus occidentalis
Blooms: April-July

Western morning glory, a native, grows in brushy areas or open woods. It is especially common in chaparral and disturbed areas.

This woody climbing vine is often found entangled in other shrubs. Its triangular leaves are alternate. Creating quite a splash of color on the brown hillsides, the large funnel-shaped flowers turn from white to purple with age.

Manzanita

Manzanita

Western Morning Glory

BLACK SAGE

Labiatae
(Mint) Family

Salvia mellifera
Blooms: March-August

Black sage is a common inhabitant of chaparral in the central and southern portion of the coastal mountains.

This shrub has oppositely arranged oblong leaves which have slightly toothed margins. The aromatic leaves, characteristic of the mint family, are a distinguishing feature. Whitish to pale blue tubular flowers grow in clusters along a tall stem, often reaching high above surrounding plants.

The seeds of black sage were used to flavor food by early Californians, who added them to cooking meats and poultry. A tea can be made by soaking these seeds in water.

PITCHER SAGE

Labiatae
(Mint) Family

Lepechinia calycina
Blooms: April-July

This perennial sage is common in chaparral communities in the central and southern coastal mountains.

Pitcher sage grows as a medium-sized shrub. The aromatic leaves are slightly toothed and covered with small hairs. Characteristic of most of the mint family, the stems are square. Single, white, tubular flowers appear in mid-spring, attracting numerous bees to their nectar.

Black Sage

Pitcher Sage

SALAL

Ericaceae
(Heath) Family

Gaultheria shallon
Blooms: April-July

Salal is a common plant of the redwood and mixed evergreen forest floors. It can often be seen hanging over trail ledges and stream and road banks.

This plant grows as a low-spreading shrub. Ranging in length from 2 to 4 inches, the thick, dark green leaves are oblong with pointed tips. The pinkish bell-shaped flowers appear in late spring and hang in rows off a central axis. With the onset of summer, berries replace the flower groups and turn black when ripe.

SOAP PLANT

Liliaceae
(Lily) Family

Chlorogalum pomeridianum
Blooms: May-June

Soap plant is found on dry slopes, often intermixed with plants of the chaparral. It is common throughout the coastal mountains.

This perennial lily grows from an underground bulb and produces a cluster of basal leaves each fall. These conspicuous leaves are long and narrow with characteristic wavy margins. In late spring, clusters of small white flowers are borne atop a tall branching stem. These delicate flowers are rarely seen by hikers, since they open only in the early evening and on cloudy days.

There are numerous uses for the soap plant. The leaves, when picked young, produce a green dye that California Indians used in tattooing and are edible either raw or cooked. Uncooked bulbs, containing saponin, a lather-producing substance, can be crushed and used for shampoo and soap. Indians baked the starchy bulbs and ate them like potatoes. When cooked slowly, they lost their soapiness and were quite nutritious. As the bulb were roasted, a thick substance was exuded which was collected and used as glue to attach feathers to hunting arrows.

Salal

Soap Plant

Soap Plant

REIN ORCHID

Orchidaceae
(Orchid) Family

Habenaria unalascensis
Blooms: April-September

Rein orchids grow on fairly dry soils, often within the mixed evergreen and oak woodlands.

These small plants grow as simple, erect stalks up to 12 inches high. Linear leaves, usually 2 to 4 in number, arise from the stem base. Covering the top of the spike in tightly packed clusters are greenish-white spurred flowers.

Either raw or cooked, the tuberous roots are edible. However, since these orchids are not very widespread, they should not be disturbed.

PHANTOM ORCHID

Orchidaceae
(Orchid) Family

Cephalanthera austinae
Blooms: May

This orchid inhabits dry woods along the northern coastal mountains but is extremely rare. It grows in only a few locations in the Santa Cruz Mountains.

The phantom orchid, a saprophyte, obtains its nutrients from dead organic matter in the soil. Growing as a nearly leafless stalk, this white plant reaches 10 to 15 inches in height. The few leaves which are present on the lower part of the stem are reduced to sheaths. Clusters of 2-lipped flowers, flecked with yellow centers, attach closely to the stem.

Because of this plant's ghostly appearance on the forest floor, it is aptly named the phantom orchid.

Rein Orchid

Phantom Orchid

WESTERN WAKE ROBIN

Convallariaceae
(Lily-of-the-Valley) Family

Trillium ovatum
Blooms: March-May

The wake robin is common in shady redwood forests. It grows especially well in valleys along the many mountain streams.

This small plant is characterized by a whorl of 3 large, dark green leaves atop a 5 to 8 inch stem. On a short, slender stalk above the leaves is a small, 3-petaled flower, which changes from white to purple as it ages. The name *Trillium* refers to the fact that the leaves and flower parts are in threes.

The thick, fleshy, underground stems cause violent vomiting when eaten.

HAIRY STAR TULIP

Liliaceae
(Lily) Family

Calochortus tolmiei
Blooms: April-June

This member of the lily family is common in open sunny locations, often inhabiting roadside embankments or lightly wooded hillsides.

The hairy star tulip has long, linear leaves and, like other lilies, grows from an underground bulb. Growing on stems 4 to 8 inches long, the beautiful whitish flowers have petals tinged with lavender in the center and covered with soft, bristly hairs. These characteristic hairs also give it the common name of "pussy ears." Three-lobed seed capsules replace the flowers when mature.

The scientific name, *Calochortus*, comes from the Greek words *kalos*, meaning beautiful, and *chortos*, meaning grass. It refers to the beauty of the flowers and the grass-like appearance of the leaves.

Western Wake Robin

Hairy Star Tulip

SUGAR-SCOOP

Saxifragaceae
(Saxifrage) Family

Tiarella unifoliata
Blooms: May-July

These small plants grow in shaded areas, often along streams and moist slopes.

Sugar-scoops have deeply lobed, long stemmed, basal leaves with toothed margins. The leaves are covered with small hairs, giving them a soft texture. Small white flowers grow in narrow panicles along a central stem, later maturing into capsules. The appearance of these capsules resembles the shape of a sugar scoop, hence the common name.

Numerous other members of the Saxifrage family have similar leaf and flower characteristics. Alum root, fringe cups, brookfoam, woodland star and California saxifrage are difficult to distinguish without a detailed key, but are also quite common in the coastal mountains.

TWO-EYED VIOLET

Violaceae
(Violet) Family

Viola ocellata
Blooms: March-June

The two-eyed violet is common on shaded slopes in mixed evergreen forests throughout the coastal mountains.

Growing from a thick, fleshy, underground root stock, this plant reaches 5 to 10 inches in height. It has bright green, heart-shaped leaves which grow either basally or along short branching stems. The delicate flowers have 5 petals, the lower 3 white and the upper 2 white on the inside and purple on the outside. Purple veins line the lower central petal, and the other two lower petals each have single purple spots. These spots, or eyes, give this plant its common name.

Violets have nectar at the base of each flower. The purple veins and spots on the petals serve to attract bee pollinators to the nectar supply. Like its relative, the redwood violet, the leaves of these plants are high in vitamins A and C.

Sugar-scoop

Two Eyed Violet

POISON OAK

Anacardiaceae
(Sumac) Family

Toxicodendron diversilobum
Blooms: March-May

This well-known plant grows in a great variety of habitats throughout the coastal mountains. It is equally at home from dry chaparral to moist, shady redwood forests.

Poison oak has many diverse growth forms, ranging from bushy shrubs to long vines. These vines can have extremely stout stems several inches in diameter and can grow up the sides of trees 40 feet tall or more. All variations have 3-lobed leaves, small whitish flowers, and white berries in summer. They all have skin-irritating oils in the leaves and stems.

In spite of its poison, this plant has many cultural uses. Early Indians, being largely unaffected by the poison, used the stems for thread, warp, and foundation in their baskets, and the juice to dye weaving material black. They also were known to draw patterns on their faces with poison oak juice, then tattoo them with a sooty California nutmeg needle, thereby getting an unfading bluish-green tattoo. Medicinally, the juice from the stems, leaves, and roots was used as a cure for warts and ringworm, and as an antidote for rattlesnake venom.

THIMBLEBERRY

Rosaceae
(Rose) Family

Rubus parviflorus var.*velutinus*
Blooms: March-August

The thimbleberry is extremely common in moist redwood and mixed evergreen forests.

Thimbleberry shrubs can reach 3 to 4 feet tall, forming dense thickets. Its large palmate leaves are covered with soft woolly hairs. Simple, rose-like flowers appear in spring, later ripening to hollow berries. These thimble-shaped berries give the plant its name.

As with many other berries, these fruits are edible both raw or cooked into jams and pies. Instead of using rouge, pioneer women reddened their cheeks by rubbing them with soft, hairy thimbleberry leaves.

Poison Oak

Poison Oak

Thimbleberry

Thimbleberry

BANEBERRY

Ranunculaceae
(Buttercup) Family

Actaea rubra
Blooms: May-July

This small perennial shrub grows along streams and in moist woods from San Luis Obispo County north to Alaska. It is only occasionally seen in Santa Cruz county, at Castle Rock State Park.

On the upper tips of sturdy stems, many flowers bloom together in a large raceme. The 3 to 5 sepals and 4 to 5 petals drop off shortly after blooming to expose several whitish stamens. This gives the raceme the appearance of a bottle brush. Although the baneberry actually has few leaves, they are pinnately compound, with many sharply toothed leaflets. The large red, or occasionally pearly white, berries mature in the summer months.

These berries are poisonous and are closely related to European species that have been responsible for the fatal poisoning of children.

TOYON

Malaceae
(Apple) Family

Heteromeles arbutifolia
Blooms: June-July

The toyon or Christmas berry shrub is common in chaparral and as an understory in open wooded forests. It is also the official shrub of the State of California.

This native evergreen shrub is sometimes mistaken for a young tan oak tree. Like the tan oak, the leaves are glossy green, brittle, and toothed. However, a closer look shows that toyon leaves have veins which diverge before reaching the margin and are more finely-toothed than tan oak leaves. In early summer, clusters of drooping white flowers appear, then in late November and early December they ripen into large clusters of bright red berries.

Like the Spanish, the Indians rarely ate the berries raw due to their bitter taste when fresh. Instead, they either hung branches over hot coals or tossed individual berries in a cooking basket along with coals.

Baneberry

Toyon

Toyon

WOOD STRAWBERRY

Rosaceae
(Rose) Family

Fragaria californica
Blooms: January-June

Wild strawberries are often found in shaded woods or near streams.

These small plants can propagate by growing runners which take root and grow more plants. Their leaves, composed of 3-toothed leaflets, arise directly from the base. White, rose-like flowers grow singly on separate stems. In early summer, these delicate blooms ripen into small scarlet berries.

The berries are juicy and delicious both raw or cooked into preserves and pies. They are high in vitamin C, thiamine, niacin, and riboflavin. Some Indians also used the green leaves in tea for the treatment of dysentery.

SLIM SOLOMON'S SEAL

Convallariaceae
(Lily-of-the-Valley) Family

Smilacina stellata var.*sessifolia*
Blooms: February-March

This herb is common in shady mixed evergreen and redwood forests.

Although it has the same growth form as fat Solomon's seal, slim Solomon's seal has longer, narrower leaves which don't clasp the stem. It also has fewer flowers, which grow in clusters off the tip of the stem. These blossoms mature into green berries striped with red.

The unusual common name was given to this plant because it resembles the East Coast Solomon's seal whose flowers mimic the mystic double-triangle symbol called Solomon's seal.

The genus name *Smilacina* means scraper and refers to the hairy stems of some species.

Wood Strawberry

Slim Solomon's Seal

Slim Solomon's Seal

FAT SOLOMON'S SEAL

Convallariaceae
(Lily-of-the-Valley) Family

Smilacina racemosa var.*amplexicaulis*
Blooms: March-May

Fat Solomon's seal is common in shady mixed evergreen and redwood forests.

The large oval leaves of this herb are pointed at the tip and clasp around the distinct, unbranched stem. Growing on tiny stalks off the main stem, numerous white flowers cover the tips of the stems for 2 to 4 inches. In summer, these flowers ripen into bright scarlet berries.

These berries are edible and tasty, but somewhat cathartic. After being soaked in lye to remove the bitterness, then parboiled to remove the lye, the starchy rootstocks can be eaten; young shoots can also be boiled and eaten.

STAR LILY

Melanthaceae
(Bunch Flower) Family

Zygadenus fremontii
Blooms: February-September

Usually found on grassy or brushy slopes, the star lily is most conspicuous after a fire. It also grows on dry ridges under madrone and tan oak forests.

Star lily is a bulbous plant with narrow linear leaves growing from the base. Its star-shaped white flowers grow in a cluster on top of a tall stalk, which often reaches to 1 or 2 feet in height.

This plant is related to the death camas, which has an extremely poisonous bulb.

It receives it genus name, *Zygadenus*, meaning join-gland, from the small gland at the base of each petal. Its species name is in honor of John C. Fremont, a famous pioneer explorer.

Fat Solomon's Seal

Fat Solomon's Seal

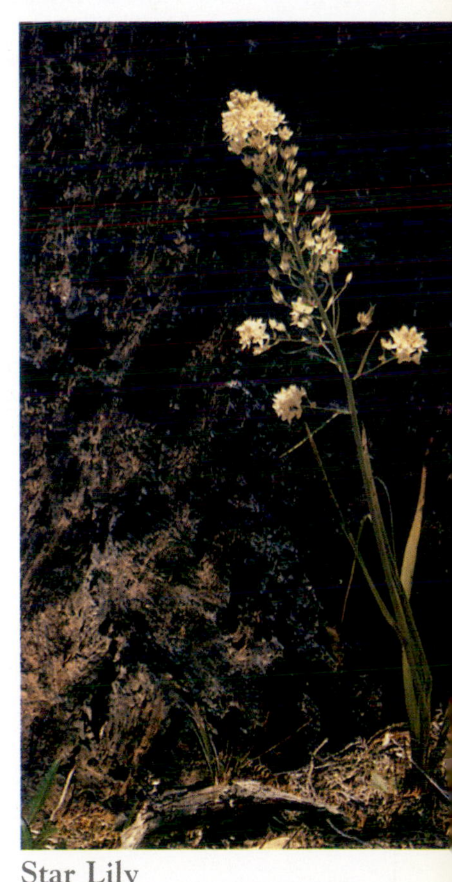

Star Lily

COMMON YARROW

Compositae
(Sunflower) Family

Achillea millefolium
Blooms: April-July

As its name indicates, common yarrow is abundant in all dry areas.

A dense cluster of tiny white flowers grows on the top of a slender 1-1/2 to 3 foot tall stalk. Soft, feather-like basal leaves grow to be about 8 inches long.

Used by ancient people to treat colds, fevers, and other ailments, yarrow has been known for centuries as a remedy. However, care should be taken in its use since it may contain some alkaloid poisons.

Its genus name *Achillea* is for Achilles, who, it was said, used a species of yarrow to treat the wounds of his warriors. The species name *millefolium* means thousand-leaved and refers to the very finely-divided basal leaves.

YERBA DE SELVA

Hydrangeaceae
(Hydrangea) Family

Whipplea modesta
Blooms: March-June

Yerba de selva is common in shady redwood and mixed evergreen forests. However, it is somewhat inconspicuous except when blooming.

This dainty plant has such a tiny stem that it can usually be found drooping almost to the ground. At the end of the stem is a cluster of tiny white flowers. Leaves are about an inch and a half long and very slender.

Its Spanish name means "weed of the forest," while its genus name is in honor of Lieutenant Whipple, commander of an 1850s Pacific Railroad expedition.

Common Yarrow

Yerba de Selva

DODDER

Cuscutaceae
(Dodder) Family

Cuscuta sp.
Blooms: May-November

There are over 15 species of dodder in California, several of which inhabit the redwood region. Growing as a parasite, some species of dodder are specific to one host while others are more general. Often the community in which the dodder is found can reflect the range of the host species. In the Santa Cruz Mountains, canyon dodder (C. *subinclusa)* is the most common and is often found on chaparral plants.

Like other parasites, the dodder plant has no chlorophyll to produce its own food and must derive its nourishment from other plants. The bright yellow-orange stems create splashes of color on the hillsides as they drape themselves over their hosts. Although the vine does have leaves, they have been reduced to minute scales. The small, waxy-white flowers are bell or urn shaped, and grow in compact, flat-topped clusters.

Dodder has a unique lifestyle. From mature flowers, one to four seeds are released and sprout in the ground. As the slender vine emerges from the ground it immediately seeks a host plant for nourishment. Upon contact with its "victim," the dodder vine climbs and attaches itself to the host with small sucker-like organs called haustoria, a name that means "to drink" in Latin. When well attached, the vine from the ground dries up and disappears like a no longer needed umbilical cord.

Like any successful parasite, dodder depends on the continued vitality of its host plant for survival. Sometimes, under harsh conditions where two or more dodder plants live on a single plant, the host can be killed. Usually, however, a dodder plant will spread itself over several host plants, enabling all to survive.

Dodder

Dodder

INDIAN PIPES

Monotropaceae
(Indian Pipe) Family

Monotropa uniflora
Blooms: June-August

This bizarre looking plant grows in damp, shady forests from Humboldt County north.

Shaped like a shepherd's hook, each 2 to 10 inch stem supports a single flower. Often these stems grow together in small clumps. The entire plant is waxy white, blackening with age. Small, scalelike leaves are practically transparent, as are the 5 to 6 petals on the drooping pipe-shaped flower. A similar species, pine sap (*Monotropa hypopitys*) is red with multiple flowers on each stem.

Lacking chlorophyll to produce their own food, indian pipes have developed an unusual method of obtaining nutrients. Instead of absorbing nutrients from duff as was once believed, these plants actually team up with a fungus to parasitize green plants. The fungus acts as a "bridge," robbing the roots of the green plants of nutrients and passing them on to the indian pipes.

RATTLESNAKE PLANTAIN

Orchidaceae
(Orchid) Family

Goodyera oblongifolia
Blooms: July-August

This member of the orchid family grows on the dry forest floors of the coastal mountains. Not commonly found, it is a special treat for the adventurous hiker.

Arising from a fleshy rootstock, this low plant has green basal leaves. The flowers, which are white with small glandular hairs, sit in straight, terminal clusters upon a single erect stalk. There are 3 petals; 2 are alike and resemble the 3 sepals, the third forms an owl-shaped beak, with no spur. An outstanding characteristic is the white central vein on the leaves.

Rattlesnake plantain was named after John Goodyear, a 17th-century English botanist.

Indian Pipes

Rattlesnake Plantain

Rattlesnake Plantain

LABRADOR TEA

Ericaceae
(Heath) Family

Ledum glandulosum
Blooms: June-August

This stiff, many-branched shrub inhabits wet, boggy regions from Santa Cruz County to Del Norte County.

Growing 2 to 7 feet tall, this shrub has dense, resinous, and fragrant foliage. The individual leaves are leathery with a dark green upper surface and a whitish, felt-like lower surface. Above the leaves, white rose-shaped flowers bloom in large rounded clusters. Like the rhododendron, to which it is related, the Labrador tea flowers mature from large scaly buds. Many tiny winged seeds are released in late summer.

The foliage of this plant is poisonous.

WESTERN LADIES' TRESSES

Orchidaceae
(Orchid) Family

Spiranthes porrifolia
Blooms: July-August

Growing in wet, boggy areas, western ladies' tresses inhabit the Coast Range from the Santa Cruz Mountains to the state of Washington.

This plant grows from tuberous roots and produces a cluster of basal linear leaves. A single stalk rises 4 to 20 inches in height and is covered with spiraling rows of small white-yellow flowers. The flowers have 3 projecting sepals, which surround 3 petals. Two petals fuse to form the upper hoods and one becomes a triangular-shaped lower lip.

Pollination of the ladies' tresses is usually done by bees. Pollen is deposited on the bee as it flicks its tongue down the narrow passage to reach the nectar at the base of the flower. This pollen can now be transported to another blossom.

The genus name is Greek and refers to the spiralling inflorescence.

Labrador Tea

Western Ladies' Tresses

CALIFORNIA LADY'S SLIPPER

Orchidaceae
(Orchid) Family

Cypripedium californicum
Blooms: May-June

California lady's slipper inhabits wet rocky banks in mixed evergreen forests from Mendocino County to southern Oregon.

Rising in clumps from short rootstocks, unbranched stems grow 1 to 4 feet tall. Linear, parallel-veined leaves clasp these stems at intervals. On the upper reaches of the plant grow 7 to 10 solitary flowers originating from the leaf axils. Three greenish-yellow sepals, one above and two fused below, surround the puffy pouch or "slipper" formed from the petals. Usually white, the pouch can also be tinged with pink or spotted with brown. On its upper surface near the base is a small opening.

Attracted to the sweet nectar, small bees enter through this opening, but are unable to escape because of its inward turned edges and slippery sides. By following translucent "road signs" and gaining footholds on small hairs, the insect is guided to two smaller openings near the stamens where it can escape. Before squeezing out of these openings the bee is dusted with sticky pollen. This one-way "traffic pattern" ensures that the insect first passes the stigma, and then the anthers, preventing self-pollination of the plant. Unbelievably, the lady's slipper is one of the least specialized of the orchids.

FALSE LILY-OF-THE-VALLEY

Liliaceae
(Lily) Family

Maianthemum dilatatum
Blooms: May-June

Growing in low patches, false lily-of-the valley inhabits moist redwood and mixed evergreen forests along the north coast.

Six to fourteen inches tall, this is a short plant which spreads from underground roots. Its 2 to 4 leaves are heart-shaped, deeply cleft at the base, and parallel veined. Small delicate racemes of tiny white flowers perch above the large leaves on separate stalks. Each flower has 4 petals. The tan-red berries ripen in mid-summer.

The plant's scientific name is Greek for May flower.

California Lady's Slipper

False Lily-of-the Valley

SUNDEW

Drosera rotundifolia

Droseraceae
(Sundew) Family

Blooms: July-August

This perennial herb grows in cold acidic bogs and swampy areas of mixed evergreen and redwood forests from northern California to Alaska.

Sundew, a carnivorous plant, has rounded basal leaves covered with sensitive red hairs. These hairs secrete a gelatinous fluid which entraps insects. Clusters of white-pinkish flowers sit atop a naked stalk. There are usually 5 petals, surrounded by 5 united sepals.

Insects that land on the leaves activate the sensitive hairs, causing the leaf to fold inward and moving the animal to the center of the leaf. Here, in this digestive area, the insects are "eaten." The nutrients are extracted and used by the plant for food.

WINDFLOWER

Anemone deltoidea

Ranunculaceae
(Buttercup) Family

Blooms: April-May

The windflower grows on moist wooded slopes below 5,000 feet along the coast of northern California.

Growing from slender, creeping rootstocks, this plant can be 8 to 22 inches in height. The toothed leaves are divided into 3 leaflets and adhere tightly to the delicate stem. Reaching above the leaves is the solitary white flower. Unlike most flowers, the windflower has no petals. The 5 sepals, green in most plants, are white and take on the appearance of true petals. There are numerous yellowish-green stamens surrounding the central pistil.

The genus name comes from the Greek word *anemos*, meaning wind.

Sundew

Windflower

VANILLA LEAF

Berberidaceae
(Barberry) Family

Achlys triphylla

Blooms: April-June

Usually growing in patches from creeping rootstocks, this herb can be found in moist redwood and mixed evergreen forests from Mendocino to British Columbia.

Two slender stalks grow side by side during the blooming season. One, the leaf petiole, is topped by a large round leaf composed of 3 fan-shaped, deeply notched leaflets. The second stalk is taller and supports a 1 to 2 inch long terminal spike of flowers. Having no petals or sepals, these small flowers are composed of 6 to 13 white stamens surrounding a central ovary. The scientific name, *Achlys*, comes from Achilles, the Greek god of the night.

PEARLY EVERLASTING

Compositae
(Sunflower) Family

Anaphalis margaritacea

Blooms: May-September

Pearly everlasting is commonly found in open, dry fields and meadows and along roadsides. It does well in poor soil and can often be seen in disturbed areas throughout the Coast Range.

At first glance, the rounded clusters atop grayish woolly stems seem to be composed of pearly white flowers with brownish-yellow centers. A closer look shows that the dark centers are the actual flowers, which are surrounded by papery white bracts. This perennial grows about 2 feet tall, with narrow, dull-colored leaves 3 to 4 inches long.

Pearly everlasting is one of those plants which can usually be found by following your nose. It smells strongly of maple syrup or curry, depending on your particular tastes and probably which meal you're looking forward to! Papery even when young, the flower heads are often used in dried flower arrangements.

Similar to the everlasting is the genus *Gnaphalium* or cudweed. The papery flowers are in compact heads arising from a woolly stalk. These, too, are very fragrant.

Vanilla Leaf

Pearly Everlasting

COYOTE BRUSH

Compositae
(Sunflower) Family

Baccharis pilularis ssp.*consanguinea*

Blooms: August-October

Coyote brush is a common shrub of chaparral, the edges of open woods, and disturbed areas. Several species are native to California.

These evergreen plants grow 2 to 5 feet in height and width. The small leaves, 1/2 to 1 inch long, are oblong and coarsely toothed. Whitish-yellowish flower clusters form at the ends of the branches and branchlets in leafy panicles. They are dioecious; male and female flowers form on separate shrubs.

The word *Baccharis* comes from the Greek god *Bacchus*.

Coyote Brush

COMMON SNOWBERRY

Symphoricarpos albus

Caprifoliaceae
(Honeysuckle) Family

Blooms: May-July

This small shrub is often found hidden in the undergrowth of shaded mixed evergreen and oak woodlands.

With slender stems and small oval leaves, this plant is inconspicuous most of the year. Its rose-pink flowers grow on short-stemmed racemes. With the coming of summer, the small white berries, which give the plant its name, appear and make identification easier. White-berried plants are unusual in this area, but watch out, don't confuse them with the poison oak's greenish-white berries!

Also common in these mountains is the creeping snowberry, *S. mollis*. A trailing shrub/vine, the stems and leaves are covered with small hairs and the berries are smaller than those of the common snowberry.

Some Indian groups used the lightweight snowberry as a raw material for arrow shafts. They would cut the bushes back in the fall so that straight shoots would grow the following spring. The pith was drawn out to make the desired weight shaft. Although most berries of this species are edible either raw or cooked, some evidence suggests that the poisonous drug, saponin, may be present in *S. racemosus*, a species found from Lake County north. Saponin is also found in the leaves, making them inedible. Indians were known to use the roots as a treatment for colds and stomach aches by steeping them into a tea.

Snowberry is also an important plant for wildlife; bees produce a white honey from the nectar and pollen, animals eat the berries, and birds use the bushes for shelter.

Common Snowberry

WILD CUCUMBER

Marah sp.

Cucurbitaceae
(Gourd) Family

Blooms: March-May

Two similar species of wild cucumber are found in the coastal mountains. M. *oreganus* grows in the more northern region, while M. *fabaceus* grows in the more southern areas.

These climbing vines grow so profusely that they often obliterate the shrubs and small trees over which they spread. Numerous tendrils and large grape-like leaves make this an easily recognizable plant. Although they grow on the same vine, male and female flowers are distinct. Male flowers grow on branching racemes; female flowers grow as individuals. Flowers of M. *fabaceus* are greenish-cream and cup-shaped. Those of M. *oreganus* are white and more bell-shaped. The fruit is a large, spiny burr with four round seeds enclosed.

This plant earned the name "old man in the ground" and "manroot" because the roots are huge, often the size and vaguely the shape of a human. The round seeds were used as marbles by children of Spanish settlers.

The seeds from the wild cucumber were used for a variety of ailments. The large, brown seeds, extracted from the pod and covered with a soapy pulp, were roasted and eaten for kidney trouble. Oil from the seeds was used to cure falling of the hair. The roots, bitter if tasted, were mixed with sugar and used to treat saddle sores on horses.

Wild Cucumber

YERBA BUENA

Labiatae
(Mint) Family

Satureja douglasii

Blooms: May-August

Sometimes found in open areas, but more common in mixed evergreen or redwood forests, yerba buena grows from British Columbia south to California.

A tiny perennial with stems only 1/2 to 3 inches long, this plant is usually found by smell rather than sight. When crushed, its strong mint aroma permeates the surrounding area. Also inconspicuous, the small white flowers grow singly from the leaf axils.

Like other mints, yerba beuna is well adapted to pollination by insects. The lower petals are fused to form a landing platform. As an insect crawls down to the nectar it rubs first on two stigmas, causing them to fold tightly together, then passes four stamens which deposit pollen on its body. To avoid self-pollination, the stigmas remain clamped together until the insect leaves the flower.

The dried leaves can be steeped 15 to 25 minutes to make a tea. It was used by California Indians and Spanish settlers to aid digestion, but unlike many medicinal teas, this one is delicious!

TRAIL PLANT

Compositae
(Sunflower) Family

Adenocaulon bicolor

Blooms: May-September

Found in the Coast Range from Santa Cruz north to British Columbia, this small perennial grows in moist mixed evergreen forests, often forming large patches.

Tiny tubular white flowers grow in terminal heads on long slender branches. The spear-shaped leaves are much more prominent. With petioles as long as the blades, these leaves are green and smooth on the upper surface, white and woolly underneath.

It is the white and woolly underside of the leaf which gives the trail plant its name. A person or animal walking through a patch inevitably breaks the long petioles and when the leaf turns over it becomes a white pointer marking the path of its disturber.

Yerba Buena

Trail Plant

FLOWERS - YELLOW TO ORANGE

BUTTERCUP

Ranunculus californicus

Ranunculaceae
(Buttercup) Family

Blooms: February-May

Buttercups are found on grassy slopes that are moist in spring. In fact, the Latin name *Ranunculus* means little frog, because both are found in this habitat.

The compound leaves are basal, lobed, and somewhat variable. The glossy, bright lemon-yellow flowers grow on long stems and are shaped like little saucers.

Indians boiled the roots like potatoes. They also tossed seeds in a basket on a windy day to clean them, then placed them in a basket with slow-burning coals, tossing them again as they roasted. Roasting removed the poisonous toxin, protoanemonim, and gave the seeds a popcorn-like flavor. After this preparation the seeds were eaten whole or ground into flour.

Western settlers pickled the young flowers. Also, a yellow dye was made by the Indians by crushing and washing the flowers.

REDWOOD VIOLET

Viola sempervirens

Violaceae
(Violet) Family

Blooms: February-June

Found in moist areas of the redwood forests, redwood violets can be seen throughout these mountains. They are especially abundant in valley floors along creeks.

The violet is a short-stemmed plant with small, heart-shaped leaves. Its tiny flowers are lemon-yellow with purple veins lining the lower 3 petals.

The species name for redwood violet, *sempervirens*, means evergreen.

Violets are related to pansies and are best known as ornamentals. They have also been used to make candied violets, violet-flavored vinegar, and violet leaf tea. However, violets are not very common in these mountains, so it would be a shame to pick them for this purpose.

Buttercup

Redwood Violet

STICKY MONKEY FLOWER

Scrophulariaceae
(Figwort) Family

Diplacus aurantiacus

Blooms: March-July

Sticky monkey flower is a hardy plant found in most dry chaparral regions in these mountains.

It is composed of woody, branched stalks with opposite, narrow, dark green leaves. These thick, sticky leaves give the plant its name. In spring the entire stalk is covered with bright orange tubular flowers. These funnel shaped flowers have two lips, with the upper one slightly longer than the lower.

Although bitter, young leaves and stems can be eaten in salads. Indians crushed the raw leaves and stems and applied them to wounds. In early spring, the flowers contain a drop of sweet nectar at the base.

SCARLET AND LARGE MONKEY FLOWER

Two other related monkey flowers are noteworthy. Both grow in moist areas along streambanks and seeps. Scarlet monkey flower (*Mimulus cardinales*) has a bright red flower and is occasionally seen. Common large monkey flower (*Mimulus guttatus*), as its name indicates, is large flowered and quite common.

Sticky Monkey Flower

Scarlet Monkey Flower

CALIFORNIA FLANNEL BUSH

Sterculiaceae
(Cacao) Family

Fremontodendron californicum

Blooms: May-June

This shrub grows on dry slopes, often on the edges of seasonal creeks.

The California flannel bush gets its common name from the fuzzy texture of its leaves. Covered with small tufts of hairs on the upper and lower surfaces, the 3-lobed leaves have a dull green appearance. Large showy yellow flowers bloom in spring, bringing a sudden surge of color to the hillsides. As the flower matures, bristly brown seed capsules form, often remaining on the branches for many months.

The bark of the flannel bush was used by early California settlers for its soothing qualities. Brewed into a tea, the mucilaginous inner bark relieved sore throats and raw membranes. This often gave it another name, slippery elm.

BUSH POPPY

Papaveraceae
(Poppy) Family

Dendromecon rigida

Blooms: April-July

Bush poppy is common in sandy or rocky soils, often in recently burned areas. It is usually associated with chaparral or closed-cone pine forests.

The bright yellow, cup-shaped flowers of this shrub are a sharp contrast with the grayish-green growth of the chaparral. Three to 4 foot tall stems are woody, with dull green, willow-like leaves.

The poppy flowers were used as a narcotic.

California Flannel Bush

Bush Poppy

CALIFORNIA POPPY

Papaveraceae
(Poppy) Family

Eschscholzia californica

Blooms: March-October

Well adapted to dry areas, the California poppy is common on chaparral slopes and along roadsides.

Each plant produces several satiny, bright orange, cup-shaped flowers, each on a separate stalk. After blooming, these flowers mature into long seed pods that contain numerous seeds. The basal leaves are dull green and are so finely divided that they appear feathery.

Because of its abundance and bright appearance, the poppy is well known as the state flower of California.

It is also known for its narcotic properties, which it shares with other members of this family. The leaves were crushed and packed around aching teeth to kill pain. Today the drug is still used in some places as a headache cure.

SKUNK CABBAGE

Araceae
(Arum) Family

Lysichiton americanum

Blooms: March-June

This herbaceous plant inhabits wet, boggy areas in the coastal mountains. Often found near springs and year-round creeks, large patches grow within the redwood community in Butano and Fall Creek State Parks. It is very common along the north coast.

Skunk cabbage is a perennial plant arising from a fleshy, horizontal root. The large, simple, basal leaves are oblong and often reach 1 to 3 feet in length. Like its relative, the ornamental calla lily, flowers are situated along a fleshy central stem, called a spadix, and surrounded by a large, single leaf, called the spathe. Both the spadix and spathe of skunk cabbage are bright yellow, contrasting beautifully with the deep green leaves. The odorous spathe gives this plant its common name.

Since patches of skunk cabbage are uncommon in these mountains, they should not be collected.

California Poppy

Skunk Cabbage

CHECKER LILY

Fritillaria lanceolata

Liliaceae
(Lily) Family

Blooms: March-May

Although it sometimes grows on dry open slopes, the checker lily is usually found in the shade of redwood and mixed evergreen forests.

When not blooming, all that can be seen of the checker lily is a large oval basal leaf. However, in early spring a 1 to 2 foot long stem grows from the small subterranean bulbs. On the upper portion of this stem are several whorls of 3 to 5 lance-shaped leaves. Unless seen together, it is hard to believe these leaves grow from the same bulb as the basal leaf. Near the tip of the stem nod greenish-yellow flowers with purple spots. Because of these bell-shaped flowers, another name for this plant is mission bells.

Since the bulbs of most species in this genus are edible either raw, boiled, or dried, the checker lily is probably no exception. Care should be taken, however, since some individuals may not be able to handle large quantities. Because of this, and the fact that these flowers are so beautiful, we recommend they not be eaten.

TIGER LILY

Lilium pardalinum

Liliaceae
(Lily) Family

Blooms: June-September

The tiger lily grows in moist shady areas of redwood and mixed evergreen forests, usually along streams. Once common in these mountains, it was uprooted and taken home by so many people that it is now rather uncommon.

The beautiful orange flowers are marked with purple spots and nod at the end of a tall stem. This stem is 2 to 3 feet tall and has several whorls of long, narrow leaves.

Legend has it that the tiger lily was created by a Korean hermit who removed an arrow from a tiger. They became friends, and when the tiger died, the hermit transformed his body into a tiger lily to preserve their friendship. Later, when the hermit drowned, the tiger lily spread down the streams looking for his lost friend.

Checker Lily

Tiger Lily

EUREKA LILY

Lilium occidentale

Liliaceae
(Lily) Family

Blooms: June-July

The eureka lily grows close to the ocean from Humboldt County north through southern Oregon. The plant is commonly found in low, wet areas growing amid ferns and thickets of the coastal scrub community.

Arising from a rhizomatous bulb, this lily grows from 2 to 6 feet in height. The dark green leaves are whorled along the central portion of the stem and are lanceolate in shape. The flowers grow atop the slender stems in clusters of 2 to 15 and nod downwards on long pedicels. The petals of the flowers curve backwards to reveal their vibrant colors—green centers yielding to yellowish-orange then dark red along the outer edges. The yellow-orange is usually dotted with maroon.

This lily is also known by the common name of western tiger lily; however, it is distinguishable from the other tiger or leopard lily (*L. pardalinum*) by the yellow central portion of the petals.

NARTHECIUM

Narthecium californicum

Liliaceae
(Lily) Family

Blooms: July-August

This lily grows in wet, boggy areas in the coniferous forests of the Coast Range, usually below 5,000 feet. It can extend itself to below 8,000 feet in the central Sierra Nevada.

Growing from a creeping rootstock, narthecium consists of a clump of linear, basal leaves and a solitary flower stalk. The small flowers are greenish-yellow and hang in a raceme from the erect stem. Six petals radiate from the receptacle, encircling six woolly stamens. Each stamen is topped with orange-red anthers that contain the pollen. When mature, the flower releases small, tailed seeds.

This plant is also known by the common name, bog-asphodel.

Eureka Lily

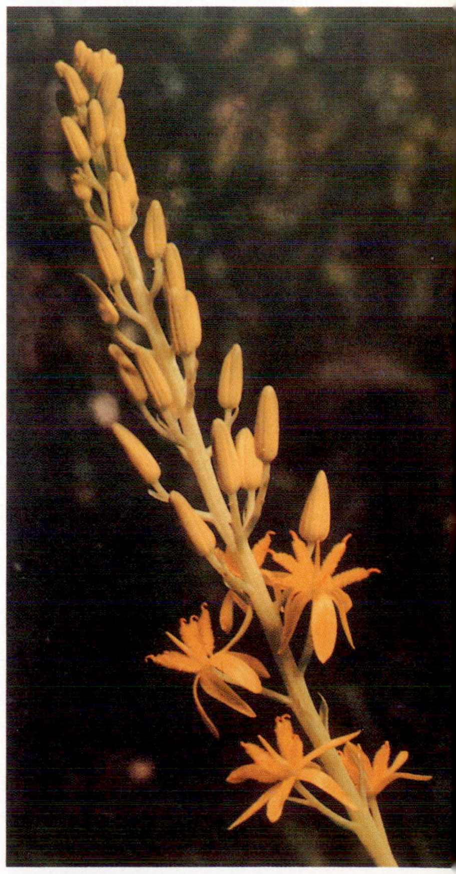

Narthecium

OREGON GRAPE

Berberidaceae
(Barberry) Family

Berberis nervosa

Blooms: March-May

Oregon grape is quite common on wooded slopes of the mixed evergreen and redwood forests in the northern coastal mountains. It is the Oregon state flower.

This erect and branching shrub has yellow wood and inner bark. The leaves are 4 to 10 inches long with 5 to 9 leaflets, somewhat resembling holly. The glossy green leaflets are leathery, with 10 to 20 spines along the margins. Showy yellow flowers grow in drooping clusters. The six sepals, which drop off early, and the six petals each grow in two whorls. The rather large blue berries have a whitish coating.

Although somewhat sour, the few-seeded berries are edible and make an excellent jelly. Medicinally, Oregon grape has had several uses. Bark was brewed into a laxative, a lotion for skin disorders, and a tonic to aid digestion.

CALIFORNIA PITCHER PLANT

Sarraceniaceae
(Pitcher Plant) Family

Darlingtonia californica

Blooms: April-June

Inhabiting bogs, marshes, and streamsides of the north coast region, the California pitcher plant is a fascinating addition to our flora. It is most common in Trinity, Del Norte, and Siskiyou counties, between 300 and 6,000 feet.

This herbaceous plant is insectivorous; its flower and leaf parts are specially developed to attract and entrap insects for its food. The greenish-yellow and deeply veined leaves are tubular at the base and gradually rise 6 to 15 inches into rounded hoods. Two lobes extend from the hood, partially covering the opening. The single flowers arise on leafless stalks; the stalks are usually taller than the leaves. Five yellowish-green sepals hang beyond the maroon-purple petals, giving the flower a nodding appearance.

Attracted to the plant by scented nectar secreted from the leaf hood, insects alight on the lobes, then enter the leaf chamber.

Oregon Grape

California Pitcher Plant

FRANCISCAN BROOMRAPE

Orobanche fasciculata

Orobanchaceae
(Broomrape) Family

Blooms: May-July

Franciscan broomrape is an occasional plant of the chaparral areas in this region. Lacking the green pigment chlorophyll, it is a root parasite and derives its nutrients from other plants.

There are several different species of broomrape, many of which are parasitic on specific hosts. The most common hosts are yerba santa (*Eriodictyon)*, chamise (*Adenostoma)*, and many other members of the composite family.

This plant grows in stout clusters, 5 to 10 inches at most in height. The scaly stem emerges from the ground for 1 to 2 inches, then bears several yellow tubular flowers, the bases sometimes tinged in purple. The flowers are composed of united sepals and 5 united petals. The upper and lower petal lips can be longer or shorter than the floral tube, depending on the subspecies.

GOLDEN FLEECE

Haplopappus arborescens

Compositae
(Sunflower) Family

Blooms: July-October

Golden fleece is commonly found in the chaparral and foothill woodland regions of the Coast Range. It is one of over 35 species of *Haplopappus* found in California.

Arising from a thick taproot, golden fleece is a stout, erect shrub growing 3 to 8 feet high. Ranging from 1 to 3 inches in length, its leaves are bright green, very narrow, and crowded along the stem. The flowers, arranged in rounded terminal clusters, are golden-yellow, hence the name golden fleece. These flowers mature to release seeds that are covered with fine silky hairs.

The name *Haplopappus* comes from the Greek—*haplos* for simple and *pappos* for pappus—and refers to the hairy seed covering.

Franciscan Broomrape

Golden Fleece

CALIFORNIA BROOM

Lotus sp.

Fabaceae
(Pea) Family

Blooms: April-June

Members of the *Lotus* genus can be found throughout the Coast Range. Although usually growing in the drier woodland or chaparral regions, a few species inhabit moist areas in the forest.

The California broom, or deerweed, is a low-growing plant. Its alternate leaves are pinnate, composed of several oblong leaflets which usually grow on short stipules. At the tips of the stems grow the flowers, either solitary or clustered. They are usually yellow, occasionally white, and often develop a reddish tinge as they age. Composed of a large reflexed petal called the "banner," two side "wings," and two lower petals fused to form a "keel," the flowers are characteristic of the pea family. The seeds are borne in a small pod, the pod often splitting open upon maturity.

Many *Lotus* species are the first plants to colonize an area after a fire or a disturbance of the soil. Their quick growth can aid in erosion control as well as in soil enrichment, through their relationship with the nitrogen producing bacteria, *Rhizobium*, in their roots.

TARWEED

Hemizonia corymbosa

Compositae
(Sunflower) Family

Blooms: May-September

Tarweed is a common plant in the redwood region. Usually seen along the edges of trail banks and roadways, it can tolerate dry, rocky soils and is a pleasant addition to our summer flora. Several different species are found in California and can inhabit virtually all plant communities. The long, slender stems and linear leaves of the coast tarweed are sweetly scented. Its yellow flower heads, sitting atop a cup-like base, have 8 to 20 outside rays with as many or more center disk flowers.

Also known by the common name of tarweed are species in the genus *Madia*. They are quite similar to *Hemizonia*, yellow flower heads atop long slender stems.

California Broom

California Broom

Tarweed

TWINBERRY

Lonicera involucrata

Caprifoliaceae
(Honeysuckle) Family

Blooms: March-April

Twinberry grows in moist areas of the coastal strand and closed cone pine communities from Santa Barbara to the Arctic.

One of the few deciduous shrubs of this region, twinberry has ovate leaves which are darker and wavy on the upper surface, pale and hairy beneath. Nestled in rosy red bracts, the tubular yellow flowers grow in pairs. These ripen into shiny black berries that are edible either fresh or dried.

YELLOW MARIPOSA LILY

Calochorutus luteus

Liliaceae
(Lily) Family

Blooms: April-June

The yellow mariposa lily is common to grasslands, foothill woodlands, and edges of mixed evergreen forests throughout the Coast Range. It prefers areas of heavy soil, usually below 2,000 feet.

A perennial herb growing from a scaly underground bulb, this lily is easily recognized by its showy yellow flowers. Situated at the top of a leafed stalk, the flowers are bowl shaped, with the three petals lightly marked with reddish-brown lines. Six stamens radiate from the center. Small glands at the base of each petal are crescent-shaped and covered with short matted hairs.

A related species, the white mariposa lily (C. *venustus*), grows in the southern Coast Range and Sierra. Usually white, but sometimes lilac or yellow, it can be distinguished by the squarish glands at the base of its petals. It is usually found on serpentine.

Named in Spanish "butterfly," the mariposa is a beautiful example of our native flora.

Twinberry

Mariposa Lily

FERNS AND FERN ALLIES

BRACKEN FERN

Pteridium aquilinum

Polypodiaceae
(Fern) Family

The bracken fern grows anywhere from dry, open slopes to moist, sheltered valleys.

In spring, the young fronds uncurl and develop into the adult plants which can be distinguished from other ferns by their highly branched form. Usually 1 to 4 feet tall, the main stem branches laterally to smaller stems from which grow highly divided, leaf-like pinnae. Conspicuous spores mature in small encasings on the undersides of the pinnae in late spring and early summer.

Young fronds are edible and can be eaten raw or steamed. They have a mucilaginous quality which makes them ideal for thickening soups. In fact, the Japanese so relished them that their government had to pass laws to prevent the fern's extinction.

Some references claim that older fronds are only poisonous when eaten in large quantities for an extended length of time. However, eating any part of this plant, except young fronds less than a foot high, is not recommended, since the shoots accumulate a vitamin B-destroying enzyme as they mature.

The roots were used to make basket patterns by some Indian tribes.

COFFEE FERN

Pellaea andromedaefolia

Polypodiaceae
(Fern) Family

Coffee fern is one of the few ferns which is found in dry, sunny areas. It is usually found in rocky locations in chaparral.

This hardy fern is generally under 10 inches in height. Small stems grow from the creeping rhizome and branch 2 to 4 times. Growing from these stems are tiny, rounded pinnae, which have a slightly reddish tinge.

The scientific name *Pellea* is a derivation of the Greek word *pellos*, meaning dusky, and refers to the appearance of the stems.

Bracken Fern

Coffee Fern

CALIFORNIA POLYPODY
Polypodium californicum

Polypodiaceae
(Fern) Family

The California polypody is fairly sun-tolerant but prefers shady locations. It is found in wooded valleys, often growing on rocks or trees.

This fern is unbranched, with the leaf-like pinnae growing directly along the main stem. The rounded tip of the pinnae is one of the distinguishing features of this fern. Also important for identification are the round clusters of unenclosed spores on the underside of the pinnae.

The California polypody derives its name from the Greek words *polys*, meaning many, and *podi*, meaning foot, because of the many knobby branches of the rhizome.

A related fern is the leather leaf fern (*P. scouleri*), which grows along the coast from Santa Cruz to Del Norte County, usually below 1,500 feet elevation.

WESTERN CHAIN FERN
Woodwardia fimbriata

Polypodiaceae
(Fern) Family

The western chain fern prefers moist, shady locations and is fairly common along many mountain streams.

This large fern consists of fronds which can reach 6 feet in height. The deeply divided pinnae grow in an orderly, opposite pattern, decreasing in size from the base of the frond to the tip. Spores, encased in a small brown flap called an indusium, grow in a chain-like pattern along the lower midrib of the pinnae.

These ferns were used in basket making. Patterns were made by either using the stems naturally or by dying them red with alder bark.

California Polypody

Western Chain Fern

FIVE-FINGER FERN

Adiantum pedatum var.*aleuticum*

Polypodiaceae
(Fern) Family

The five-finger fern grows in extremely moist areas and is usually found alongside streams.

Like the western maidenhair fern to which it is closely related, the five-finger fern has long slender black stems growing from a scaly base. Each of these stems ends in a palmate, or finger-like, pattern of approximately 5 to 7 smaller stems. These stems have frilly, elongated pinnae which grow asymmetrically. Like the maidenhair, the margins are curled under, hiding the spores.

The long stem of the five-finger fern was used in Indian basketry to make a black pattern. In some areas, these baskets had a special ceremonial purpose. They were used to hold the obsidian knives which were displayed in a jumping dance.

DEER FERN

Blechnum spicant

Polypodiaceae
(Fern) Family

The deer fern inhabits the banks of shady streams in redwood and mixed evergreen forests. Although somewhat uncommon in the Santa Cruz Mountains, large concentrations do grow along Sempervirens and Berry Creeks within Big Basin Redwoods State Park.

This medium-sized fern has small, dark-green pinnae which grow along a central stem. The erect fronds grow in two forms, either sterile or fertile. The sterile evergreen fronds, ranging from 7 to 40 inches in height, have short basal stalks with numerous broad pinnae. These pinnae are crowded along the stem and begin almost from the base of the plant. The fertile fronds are taller, with narrow pinnae growing along the upper two-thirds of a long, naked stalk. Spores develop in small clusters, called sori, and are arranged in parallel rows along the midrib of the fertile pinnae.

Five-Finger Fern

Deer Fern

GOLDENBACK FERN

Pityrogramma triangularis

Polypodiaceae
(Fern) Family

The goldenback fern often inhabits shaded spots in mixed evergreen and oak forests. Occasionally it can also be found on dry brushy slopes.

This fern has small triangular fronds which grow atop a slender black stem, ranging from 1 to 4 inches in height. As suggested by the common name, a characteristic waxy golden powder, which comes off easily when touched, covers the underside.

California Indians wove black patterns into their baskets by using the goldenback fern stem.

LADY FERN

Athyrium filix-femina

Polypodiaceae
(Fern) Family

Lady fern grows in moist shady areas, usually next to or very near a water source.

This fern is an annual, sending up new, tightly-curled fronds every year. During late spring and early summer, the small shoots develop into large, 3 to 4 foot long, lacy fronds. These fronds are twice pinnate, which means that the highly divided pinnae grow off the main stem. Its light green color and deeply fringed pinnae make this fern one of the most beautiful plants in the forest.

Goldenback Fern

Lady Fern

CALIFORNIA MAIDENHAIR

Adiantum jordanii

Polypodiaceae
(Fern) Family

California maidenhair is found in moist shaded areas, usually on wet, rocky outcrops.

Several erect to gently curving fronds grow out of a single, scaly base. The rounded pinnae grow on small stems which branch off the 1 to 2 foot long central stem. Undercurled margins of the pinnae enclose the spores.

The black stems of the maidenhair were pounded by Indians until they broke into long flat strands, which were woven into baskets as a black pattern. It was the strands of the maidenhair which made the pattern of a special hat called the squaw cap. When a woman was widowed, her hair was burned off at the neckline, then smeared with pine pitch. On top of this, she wore the squaw cap for a year as a sign of her grief.

WESTERN SWORD FERN

Polystichum munitum

Polypodiaceae
(Fern) Family

Sword fern grows in shady areas and is commonly found in sheltered canyons in the redwood and mixed evergreen forests.

Like many other ferns, sword fern has several individual fronds arising from a single base. Usually 2 to 4 feet in length, these fronds have blade-like pinnae, arranged alternately along a central stem. At the base of each of these pinnae is a small, perpendicular projection. Since this projection resembles the hilt of a sword, it easily identifies this fern. Spores are arranged in round clusters along veins on the underside of the pinnae.

California Maidenhair

Western Sword Fern

COASTAL WOOD FERN

Dryopteris arguta

Polypodiaceae
(Fern) Family

The coastal wood fern grows well in the shaded areas found within redwood or mixed evergreen forests.

This is a dark-green perennial fern that grows 8 to 20 inches in height. Growing from a short underground stem, the fronds have slender pinnae which branch off the main axis. These pinnae grow tightly together, giving a ruffled appearance. The sori, which contain the spores, are arranged in 2 rows along the lower surface and are covered by a horseshoe-shaped flap.

HORSETAIL

Equisetum ssp.

Equisetaceae
(Horsetail) Family

Horsetails are found along streams or in shaded, boggy locations.

Rising from an extensive underground rootstalk system, the straight, usually unbranched, horsetail stems are jointed and hollow between the nodes. Tiny longitudinal grooves run the length of the plant. Leaves, when present, are reduced to toothed sheaths which grow in whorls from the joints. Horsetails are dimorphic, which means that they have both sterile and fertile stems. Like ferns, they reproduce by spores rather than seeds. These spores are borne in terminal structures which resemble small cones.

Horsetails are one of the world's oldest plants. Before evolving into their small present-day form, they covered the earth in giant forests, 50 or more feet tall.

The silicaceous minerals with which horsetails are embedded are an excellent protection, discouraging most insects and animals from eating them. Livestock sometimes graze on these plants and can be poisoned by them. However, this high mineral content can be useful. There is an old story that miners checked these plants for particles of gold to see if the nearby streams were worth panning. Also, early settlers used the rough plants as a scouring medium.

Coastal Wood Fern

Horsetail

EXOTIC PLANTS

Although this book concentrates mostly on native California plants, we'd like to mention a few of the more abundant non-natives. Some of these plants are very striking, but we'd like to point out how beautiful they *aren't* in many respects.

Non-native plants have been introduced into this area through imported animal feeds and through their use as garden plants. For example, brooms are often used in landscaping because of their fast growth and resistance to drought. Other plants spread when used as decorations, such as when showy pampas grass spikes are placed on car antennas.

Often, introduced plants can't survive in their new habitat without care, due to adverse climatic conditions or other environmental factors. Some are adapted to specialized pollinators or seed dispersers and don't reproduce well. Sometimes, however, these plants adapt too well to this environment and, lacking their normal controls such as specialized parasites, disease carriers, or herbivores, they become so successful that they overgrow the native flora.

Why this concern over native plants if they can't compete successfully with these introduced plants? Why not let the best plant win?

California has a very beautiful and diversified flora. About 30% of the species grow nowhere else. When introduced plants overgrow these natives, we lose this diversity and can end up with natural areas covered with a single exotic species. This affects the natural balance of insects and animals which normally interact with the native species and can, in time, disrupt the whole ecological community.

Many counties and parks within the state have developed policies regarding certain non-native plants and have eradication programs; consult them to see how you can help.

PAMPAS GRASS

Cortaderia selloana

Gramineae
(Grass) Family

A native of Argentina, pampas grass is found throughout the Coast Range. It is common along roadsides, abandoned fields, and cultivated gardens and is encroaching at an alarming rate into natural areas and parks. An aggressive plant, it can quickly overcome native vegetation, forming an impenetrable thicket.

These grasses grow in large clumps, easily reaching 3 to 5 feet in width and 5 to 7 feet in height. The sharply serrated linear leaves are numerous and arise from the center of the plant. The flowers, white-pinkish, are borne in large terminal panicles. Most people recognize this plant from these showy plumes. Upon maturity the seeds are released into the wind and disperse quickly to inhabit new areas.

Because of its aggressive nature and to prevent its spread, many organizations urge land owners to remove this plant from their property. Flowering plumes should be carefully removed and placed into plastic bags to prevent seed dispersal, the leaves cut back to the crown, and the roots dug up and discarded.

Pampas Grass

SCOTCH BROOM

Cytisus scoparius

Fabaceae
(Pea) Family

Blooms: March-June

FRENCH BROOM

Cytisus monspessulanus

SPANISH BROOM

Spartium junceum

These brooms are found throughout the coastal region. Native to Europe (Scotch and Spanish) and the Canary Islands (French), they are quickly invading natural areas and crowding out native vegetation.

Growing as spreading shrubs, all these brooms have yellow "pea shape" flowers. Scotch broom is nearly leafless, the leaves composed of 1 to 3 large leaflets sparingly placed along the stems. The flowers are solitary and arise from the leaf axils. The seed pods have small hairs along the margins. French broom, in comparison, has numerous leaves, each composed of 3 small leaflets that adhere closely to the stem on short petioles. The flowers are in clustered racemes at the end of the side branches. The seed pods have hairs over their entire outer surface. Growing in a spreading rush-like manner, Spanish broom has few leaves along its stems. The fragrant yellow flowers hang in loose racemes from the branchlets.

GORSE

Ulex europaeus

Fabaceae
(Pea) Family

Blooms: February-July

Gorse is a native of Europe that has become increasingly more widespread along the coast of northern and central California and into British Columbia.

With sharp spiny leaves radiating from the branches, gorse grows as a dense impenetrable mass of vegetation. The yellow pea-shape flowers grow solitary or in clusters, maturing into hair-covered brown seed pods. Several seeds are released from each pod and grow easily in disturbed soil. Because of its rapid growth, this exotic quickly crowds out an area's native species. Land owners trying to eradicate this menace can cut the foliage back to the crown, then spray any new growth with Round-up, a herbicide. The cuttings, complete with thorns, flowers, and seeds, should be burned.

French Broom

Gorse

Gorse

PERIWINKLE

Vinca major

Apocynaceae
(Dogbane) Family

Blooms: March-July

Periwinkle is a perennial herbaceous plant that is native to Europe. It is commonly grown as an ornamental in gardens and has escaped into natural areas over the past several years. It prefers moist, shady areas, often growing under redwoods.

The dark green leaves are round to oval in shape and grow opposite along the stem. Purple flowers, 5-lobed and joined into a funnel-like shape at the base, are attached to the leaf axils by long, slender petioles.

The scientific name is Latin, *vincia*, which means "to bind," referring to the use of the long shoots as a stypic to stop bleeding.

BULL THISTLE

Cirsium vulgare

Compositae
(Sunflower) Family

Blooms: May-September

Commonly found in disturbed areas of any kind, the bull thistle is a vigorous exotic, easily crowding out native plants. In a few seasons, an area can become a solid mass of impenetrable thistles. A native of Europe, the bull thistle can now be found in many regions of California. This thistle grows 2 to 6 feet tall. The stem, leaves, and flower heads are covered with spines and prickles. In the first year of growth, a basal rosette of prickly green leaves appears. In the second growing season, the flowering stalk develops. Set atop a bristly base, the reddish-purple flower head releases hairy seeds when mature.

To control the spread of this exotic, the flowering stalk must be cut off at an early stage of development and the flower heads discarded in plastic bags to prevent seed dispersal. Seeds can remain dormant in the ground for several years waiting for just the right conditions to germinate, thus areas must be continually rechecked for complete eradication.

The thistle is related to the edible artichoke, an important food crop in central California.

Periwinkle

Bull Thistle

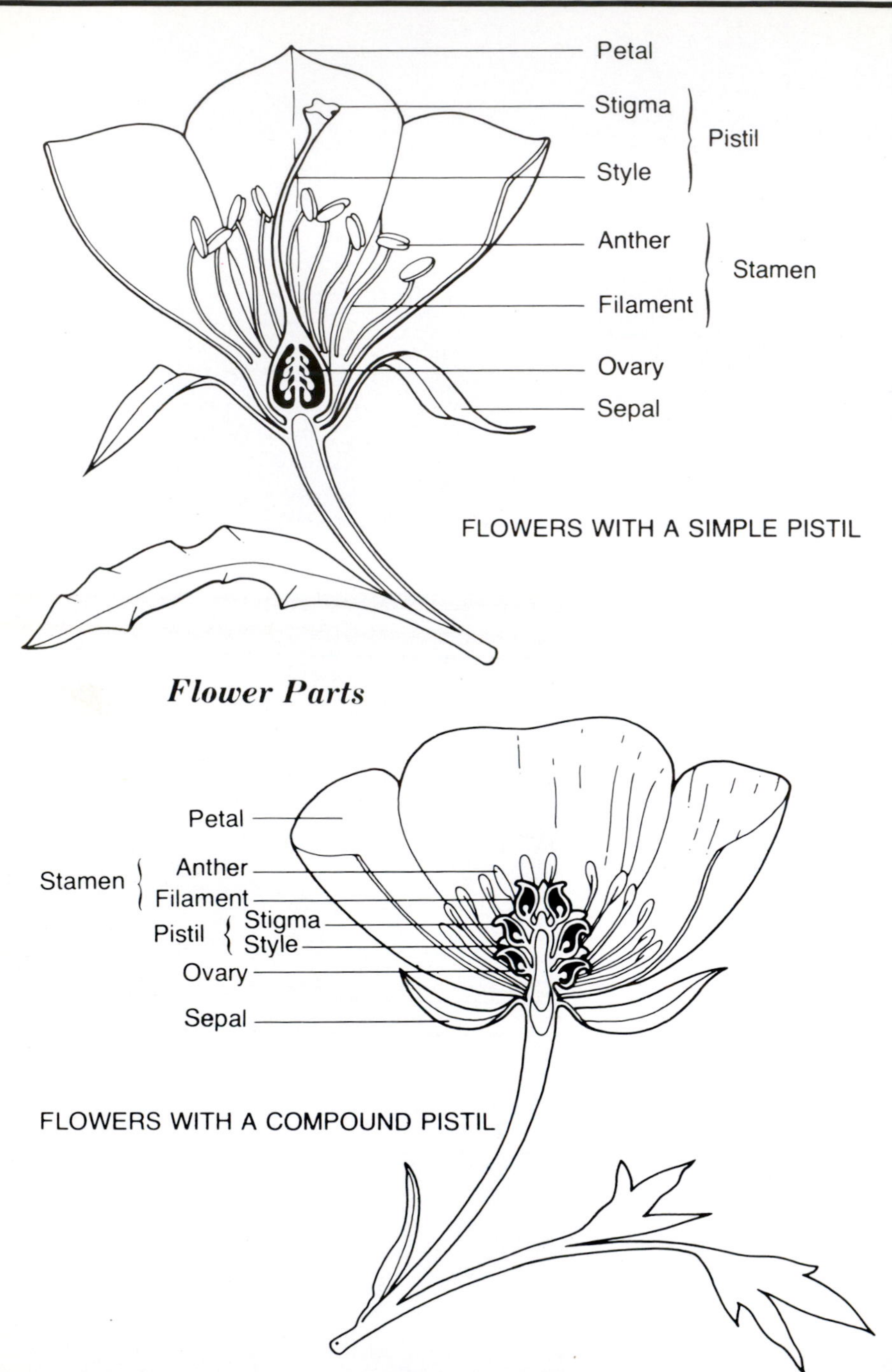
Petal
Stigma
Pistil
Style
Anther
Stamen
Filament
Ovary
Sepal
FLOWERS WITH A SIMPLE PISTIL
Flower Parts
Petal
Stamen
Anther
Filament
Pistil
Stigma
Style
Ovary
Sepal
FLOWERS WITH A COMPOUND PISTIL

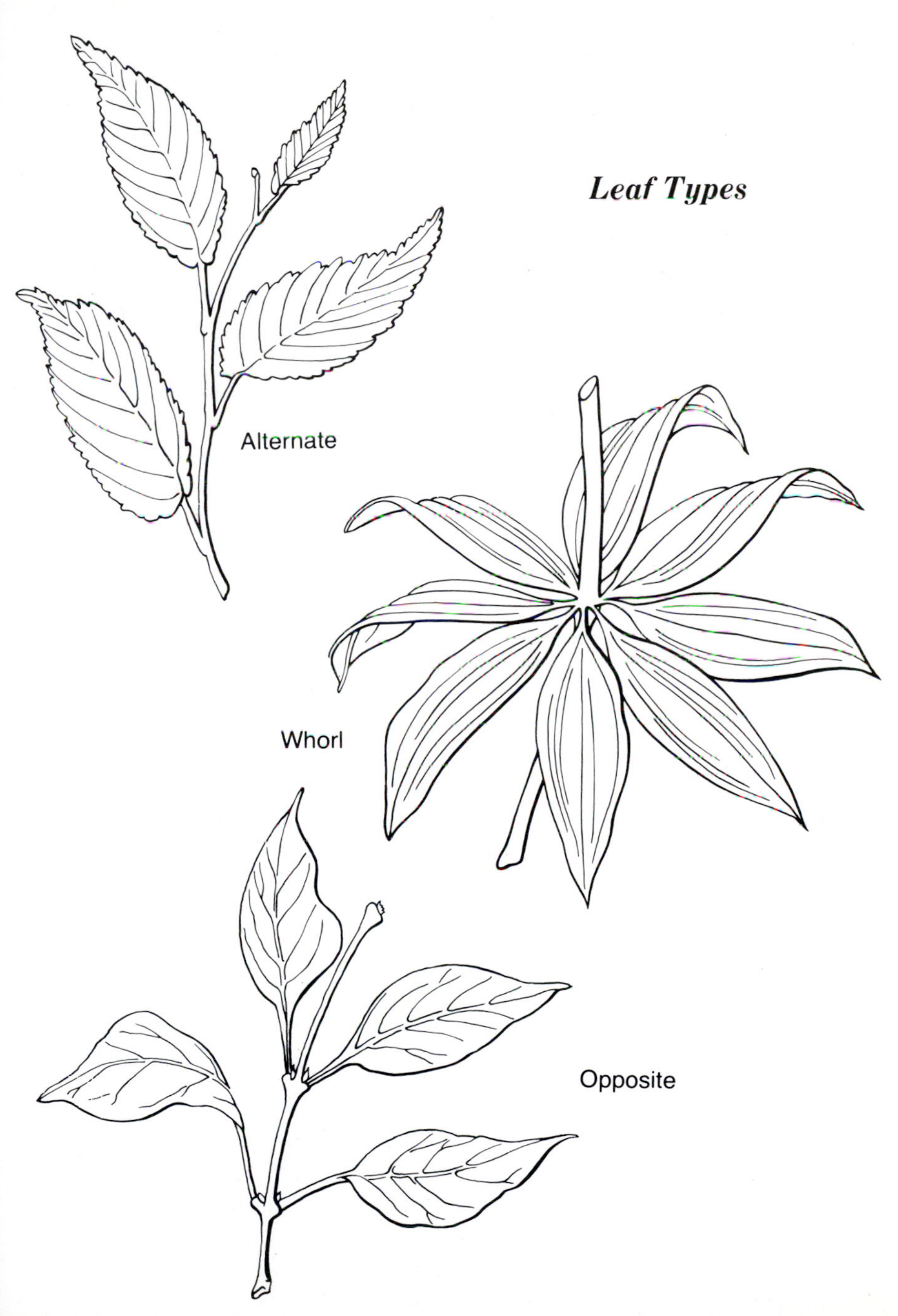
Leaf Types
Alternate
Whorl
Opposite

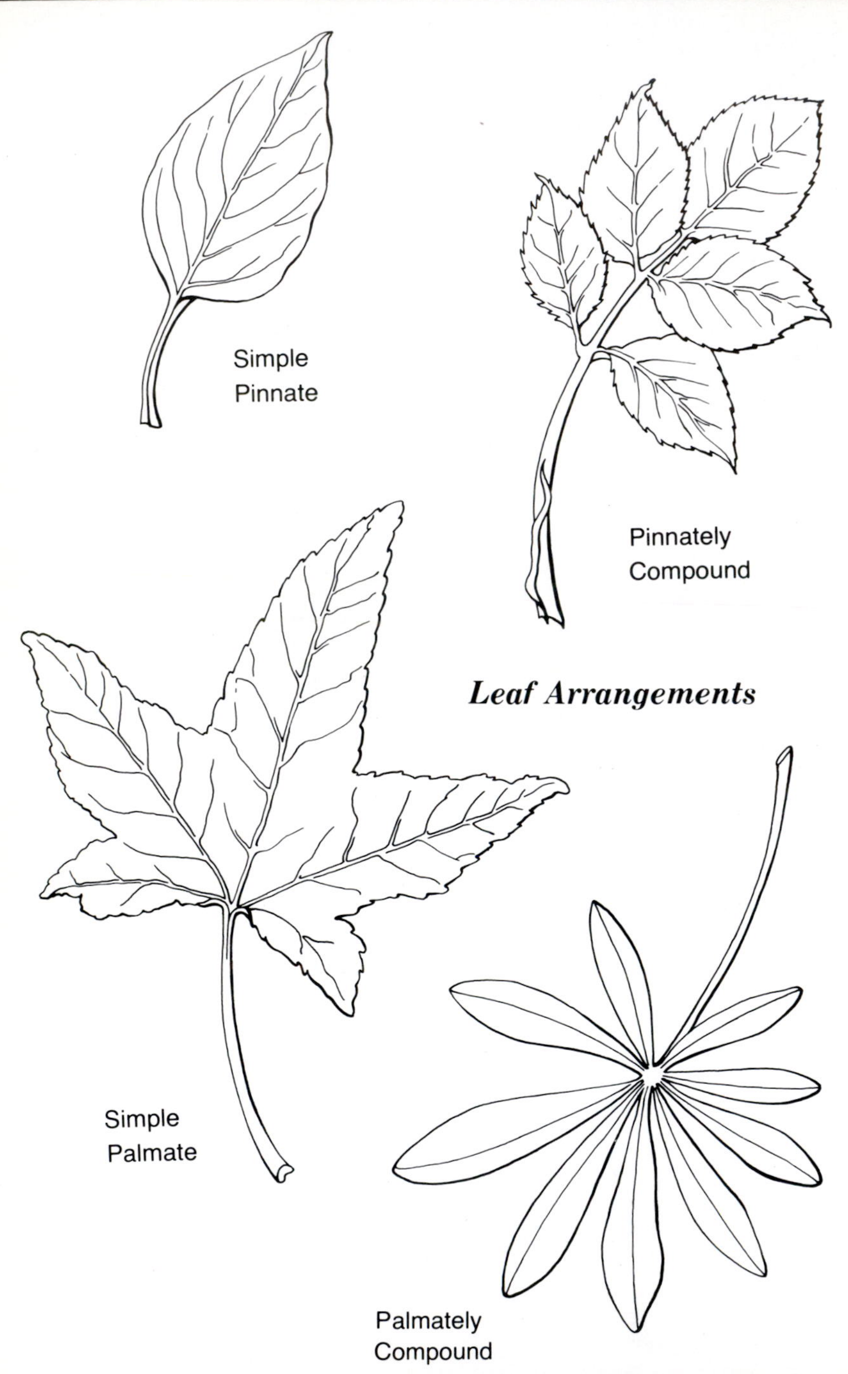

Leaf Arrangements

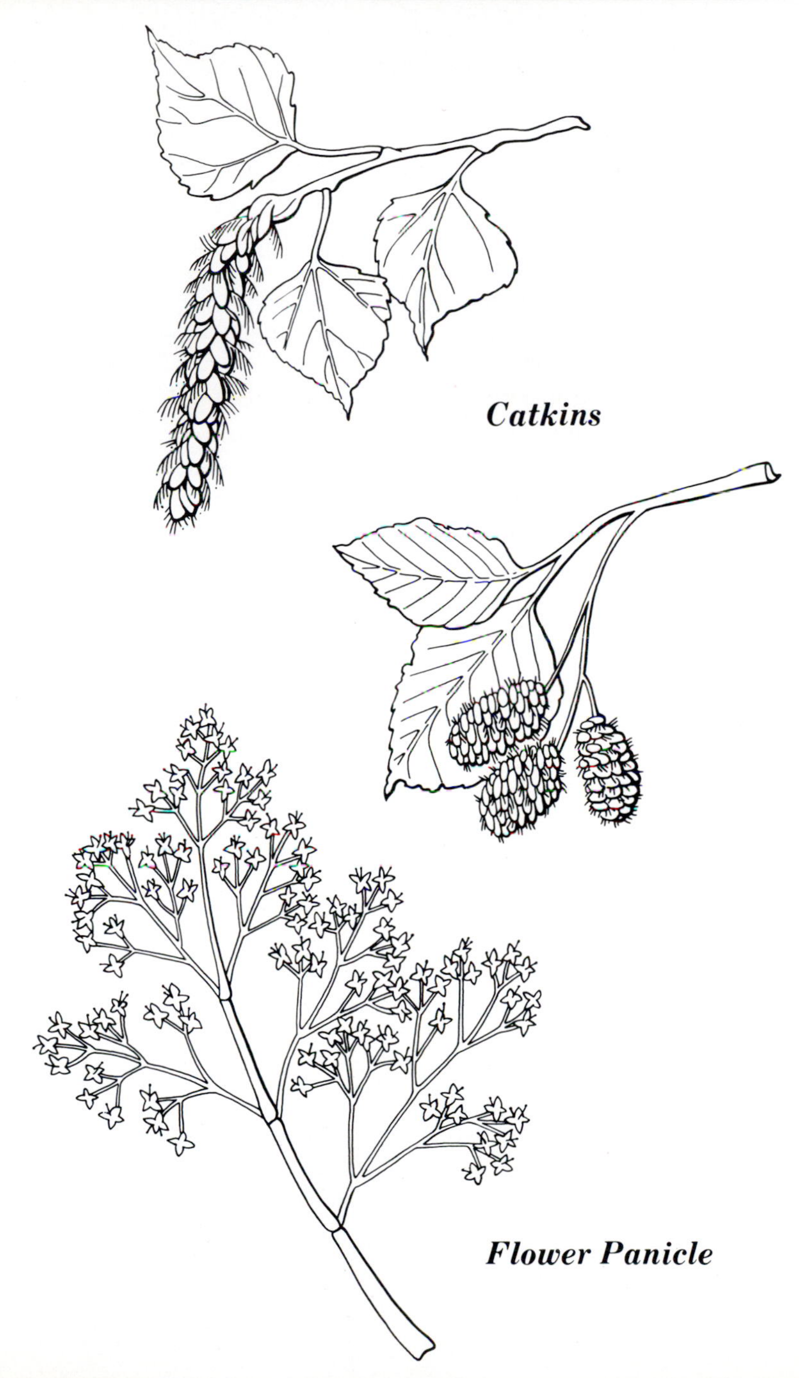
Catkins
Flower Panicle

GLOSSARY

Alternate — Leaves situated singly along a stem, arising from different leaf nodes; not in pairs.

Anther — The enlarged part of the stamen that bears pollen.

Basal — Arising from the base of a plant.

Calyx — A collective term for the sepals.

Catkin — A group of flowers, often of one sex, growing tightly-clustered along a stalk.

Compound — A leaf composed of two or more leaflets which look like true leaves.

Corolla — A collective term for petals.

Genus — A universal scientific name for a group of closely related species.

Herbeceous — Plants that do not form woody tissue.

Leaflet — A leaf-like part of a compound leaf.

Lobed — A flower or leaf that has deep indentations.

Node — The place along the stem where leaves arise.

Opposite — Leaves growing a long a stem in pairs; arising from the same node.

Palmate — Leaves with 3 or more veins developing from a common center; like a hand.

Perennial — A plant that lives from year to year, including those that die back to bulbs in winter.

Petal — Usually the showy portion of the flower parts; inside the whorl of sepals.

Pinna — (Plural, Pinnae) The leaf-like parts of a fern, often finely divided.

Pinnate — A compound leaf composed of leaflets arranged along a central stalk; featherlike.

Rhizome — A fleshy, underground stem; often called a rootstock.

Saprophyte — A plant that receives nutrients from dead organic matter.

Sepal — The outermost whorl of flower parts, usually green.

Species — A taxonomic term for closely related plants with similar morphological characteristics.

Whorl — Leaves or flower parts that grow from a single location on the stem.

REFERENCES

Abrams, Leroy and Roxana Ferris. 1960. *Illustrated Flora of the Pacific States.* Stanford University Press. Stanford, CA. Vols I–IV.

Anderson, Viola. 1980. *Flowers and Their Ancestors.* Natureweb Publications, Los Gatos, CA.

Balls, Edward. 1962. *Early Uses of California Plants.* University of California Press, Berkeley, CA.

Clarke, Charlotte Bringle. 1977 *Edible and Useful Plants of California.* University of California Press, Berkeley, CA.

Coon, Nelson. 1969. *Using Wayside Plants.* Hearthside Press, Inc., New York, NY.

Gibbons, Euell. 1974. *Stalking the Wild Asparagus.* David McKay Co., Inc., New York, NY.

Grillos, Steve. 1971. *Ferns and Fern Allies of California.* University of California Press, Berkeley, CA.

Kirk, Donald R. 1970. *Wild Edible Plants of the Western United States.* Naturegraph Publishers, Healdsburg, CA.

Merrill, Ruth Earl. 1970. *Plants Used in Basketry by the California Indians.* Acoma Books, Ramona, CA.

Metcalf, Woodbridge. 1974 *Native Trees of the San Francisco Bay Region.* University of California Press, Berkeley, CA.

McMinn, Howard and Evelyn Maino. 1947. *Illustrated Manual of Pacific Coast Trees.* University of California Press, Berkeley, CA.

Munz, Philip A., and David D. Keck. 1968. *A California Flora and Supplement.* University of California Press, Berkeley, CA.

Smith, Gladys. 1963. *Flowers and Ferns of Muir Woods.* Muir Woods Natural History Association, Muir Woods National Monument, CA.

Spellenberg, Richard. 1979. *The Audubon Society Field Guide to North American Wildflowers—Western Region.* Alfred A. Knopf, New York, NY.

Thomas, John H. 1961. *Flora of the Santa Cruz Mountains of California: A Manual of the Vascular Plants.* Stanford University Press, Stanford, CA.

Index